Protontherapy Versus Carbon Ion Therapy

BIOLOGICAL AND MEDICAL PHYSICS, BIOMEDICAL ENGINEERING

The fields of biological and medical physics and biomedical engineering are broad, multidisciplinary and dynamic. They lie at the crossroads of frontier research in physics, biology, chemistry, and medicine. The Biological and Medical Physics, Biomedical Engineering Series is intended to be comprehensive, covering a broad range of topics important to the study of the physical, chemical and biological sciences. Its goal is to provide scientists and engineers with textbooks, monographs, and reference works to address the growing need for information.

Books in the series emphasize established and emergent areas of science including molecular, membrane, and mathematical biophysics; photosynthetic energy harvesting and conversion; information processing; physical principles of genetics; sensory communications; automata networks, neural networks, and cellular automata. Equally important will be coverage of applied aspects of biological and medical physics and biomedical engineering such as molecular electronic components and devices, biosensors, medicine, imaging, physical principles of renewable energy production, advanced prostheses, and environmental control and engineering.

More information about this series at http://www.springer.com/series/3740

Marcos d'Ávila Nunes

Protontherapy Versus Carbon Ion Therapy

Advantages, Disadvantages and Similarities

Springer

Marcos d'Ávila Nunes
University of Sao Paulo (USP)
Ribeirão Preto-SP
Brazil

ISSN 1618-7210 ISSN 2197-5647 (electronic)
Biological and Medical Physics, Biomedical Engineering
ISBN 978-3-319-18982-6 ISBN 978-3-319-18983-3 (eBook)
DOI 10.1007/978-3-319-18983-3

Library of Congress Control Number: 2015940975

Springer Cham Heidelberg New York Dordrecht London

Printed on acid-free paper

Springer International Publishing AG Switzerland is part of Springer Science+Business Media (www.springer.com)

My interest is to help those who suffer
by providing useful information such
as that regarding hadron therapy,
for alleviating their suffering, their pain,
and making them happy again.
I've been through this and I was helped,
I was saved, and now I wish to help others
with a similar fate...
Blessed are those who prepare thoroughly
to help others.

Marcos d'Ávila Nunes

To my wife Maria Silvia And my daughters Alessandra, Ariane, Carla, Rafaela e Samantha

Preface

Initial Considerations

After the publication of my book *Hadron Therapy Physics and Simulations* by Springer in 2014, I was encouraged to publish another text showing the advantages, disadvantages, and similarities between protontherapy and the carbon ion therapy. Even though no carbon ion therapy centers currently exist in the United States, a large number of protontherapy centers do. Still, the topic is timely as the installation of carbon ion therapy centers is being considered.

An international interchange was established with Japan (National Institute of Radiological Sciences, NIRS) and Germany (Heidelberg Ion Beam Therapy Center, HIT) during the 2013 Joint Symposium on Carbon Ion Radiotherapy, developed by the Department of Radiation Oncology, Mayo Clinic and research collaborations between the National Institutes of Health (NIH)/HIT and Colorado State University/NIRS. As a result, $200 million in grant funding was obtained for the installation of a carbon ion therapy center at Colorado State University and the formation of a consortium in Michigan with similar proposals. After all, carbon ion therapy was born in Berkeley and was taken to Chiba, Japan, by Japanese intern researchers from Berkeley.

The success reached with this technique drove the Germans to form HIT, with a 670 ton gantry, using the intensity-controlled raster scan method, gating, local effect models, and all the available resources in Oncology, obtaining excellence in fundamental and clinic research, in association with researchers from Gesellschaft fur Schwerionenforschung in Darmstadt. This was an absolute success, with a large research field. There are currently 15 ongoing clinical trials in protontherapy and carbon ion therapy.

Carbon ion therapy is a promising technique. Certainly, the current approaches will be obsolete in the future due to the development of experimental work committed to evolving the technique and achieving important clinical results to benefit patients with cancer. Therefore, this book provides the reader with a comparative analysis between protontherapy and carbon ion therapy, thanks to a broad field of research and existing studies in the literature. It is our responsibility to expand on these achievements in order to save lives.

About This Book

This book compares and contrasts the approach, advantages, disadvantages, and indications for protontherapy and carbon ion therapy. To reach these conclusions, a broad search of the literature was performed, resulting in concise information presented here in five chapters.

Chapter 1 begins with a brief history of radiotherapy and types of radiation. Then, the attention is focused on cancer in a statistically comprehensive way with a global view. Conventional radiotherapy is still widely used, especially in developing countries that lack the financial resources to buy equipment for nonconventional radiotherapy. If the price of a cyclotron or synchrotron was close to a linac, no one would use conventional radiotherapy. However, South America, for example, has no nonconventional radiotherapy equipment, although a protontherapy center provided by Ion Beam Application (IBA) is being installed in Central America (Panama).

Chapter 2 analyzes the equipment and techniques with regard to conventional equipment, hybrid systems, more advanced models, and new equipment in development, such as the cyclinac, laser, and Dielectric wall accelerators (DWA). If the expected success is obtained, these developments will lead to a significant price drop, as well as more operational facilities. A hadron therapy simulation technique is also presented in this chapter.

Chapter 3 of this book focuses on biophysical and biological properties, which are fundamental in both experimental and clinical areas. Chapter 4 presents models for determining relative biological effectiveness. Finally, Chap. 5 discusses clinical experiences with carbon ions, covering the latest literature and presenting its advantages, disadvantages, similarities, and indications for therapy with carbon ions in comparison with protontherapy.

It is hoped that this book will provide the reader with the knowledge to analyze promising techniques using carbon ions, allowing you to draw your own conclusions. Much effort is required for the development of the technique, its comparative protocols, and clinical trials. In the future. even if the considerations presented in this book are deemed obsolete, it is my hope that this book has helped to establish new therapy centers for carbon ions and thus save lives.

Acknowledgments

My special thanks go to Prof. Dr. Ugo Amaldi, President of Tera Foundation, and Prof. Dr. Roberto Orecchia, Scientific Director of Centro Nazionale di Adroterapia Oncologica (CNAO) and the European Institute of Oncology, for giving me total health support during my liver cancer treatment, with excellent advice. They are great men working toward a noble cause: to save lives that would be cut short by cancer.

I thank Prof. Dr. José Alonso, a pioneer of hadron therapy, for providing me with excellent contacts and for always being concerned with my own liver cancer. He is a great man, looking for a way to save my life.

My thanks go to Prof. Dr. Thomas Haberer, Director of HIT (Heidelberg, Germany) and Massachusetts Institute of Technology (Cambridge, Mass., USA) for offering me the opportunity to receive carbon ion therapy for treating my liver cancer. I have a deep respect and admiration for him.

My sincerest gratitude to Dr. Harry Blom, Vice President of Springer, for his support, suggestions, and constructive criticism throughout this work—and for always helping me and being concerned with my health. He is a great man!

I also thank Dr. Mayra Castro for her support with the guidelines for the preparation of the manuscript to Springer—and above all things, for the dedication in helping me to provide the best work, while always being comprehensive, respectful, and an unsurpassed professional.

I want to thank Prof. Viktória Nyámádi, a Hungarian expert in quantum mechanics and mathematics, for her articles sent (which were very useful), enthusiasm, and support to my work. She is helping me to establish the first hadron therapy center in Brazil.

My many thanks to Prof. Dr. José Antônio *Mansur* Mendes, renowned gastro-intestinal surgeon, for his friendship and support, as well as the dissemination and acquisition of resources to deploy a hadron therapy center in Brazil, with the only purpose of saving lives with solidarity and hard work!

I thank Prof. Dr. Eugen Hug and Dr. Martin Jermann, president and secretary of Particle Therapy Co-Operative Group (PTCOG), respectively, for the acceptance of my membership and permission to use the table about particle therapy patient statistics.

Prof. Dr. Angela Bracco, Nuclear Physics European Collaboration Committee (NuPECC) Chair, showed me how pleasant a permission request can be, to use of figures and tables in my work without filling lengthy forms or requiring much time and energy—all done in a simple, direct, and pleasant way. My gratitude goes to her.

To Ion Beam Applications (IBA) goes my thanks for the permission to use figures and tables on the Cyclotron C400 and compact gantry. I also appreciate the collaboration with the implementation of a hadron therapy center in Brazil, providing us all the necessary guidance through Mr. Mauro Ferreira, Vice President of Sales of the IBA—this was very useful for our work.

My thanks to CERN (European Council for Nuclear Research); Elsevier and Copyright Clearance Center; John Wiley and Sons and Copyright Clearance Center;

Oxford University Press; National Institute of Radiological Sciences (NIRS); Centre National D'Hadrontherapiue par ions carbone; Dr. Nicolas Gaudin Head, Communications Group da International Agency for Research on Cancer, World Health Organization; Prof. Dr. Wolfgang Kurt Hermann "Pief" Panofsky (in memoriam, April 24, 1919–September 24, 2007); Prof. Dr. Hans-H.Braun; Prof. Dr. Hywel Owen, David Holder, Ranald Mackay; Prof. Dr. Lukas Opalka; Prof. Dr. Wilma Kraft Weyrather; Prof. Dr. Daniel Cussol; Prof. Dr. Thomas Friedrich; Prof. Dr. M. Scholz; and Prof. Dr. Gerhard Kraft, for permission to use figures and tables.

Finally, thanks go to Miss Michele Delbon for her translation work.

Ribeirão Preto, São Paulo, Brazil Marcos d'Ávila Nunes

Contents

About the Author

Marcos d'Ávila Nunes is Associate Professor at University of São Paulo (USP), SP, Brazil. He graduated from the Faculty of Medicine of Ribeirão Preto/USP in 1967 and did postgraduate studies in advanced theoretical physics at the School of Engineering of São Carlos/USP from 1968–69. He received his title of Doctor of Science and Associate Professor from Institute of Biomedical Sciences/USP in 1972 and 1977, respectively. Dr. Nunes completed postdoctorate studies at the Massachusetts Institute of Technology (MIT, Harvard, Cambridge, 1974/75), Boston University Medical Center (BUMC, 1974/75), National Institutes of Health (NIH, 1976), American National Red Cross (ANRC, 1976), University of South Carolina (USC, 1981/82), and Wayne State University (WSU, 1982). He is known worldwide for his electrophysiological studies in biological membranes, application of thermodynamics of irreversible processes in membranes, and the development of electron paramagnetic resonance to study the behavior of molecular components of the biological membrane, subjected to a drop/rise in temperature and their simulations (molecular cryobiology–spin labeling). He has also published work in medical informatics and participated in national and international congresses (in the position of chairman) on molecular biophysics. Dr. Nunes taught seven courses in postgraduate studies at USP. After retiring from USP as Associate Professor, he decided to devote himself to writing books in nuclear medicine (hadron therapy) and advanced physics (large hadron collider). For additional information, see: http://marcosnunes.page.tl.

Acronyms and Abbreviations

CERN	Conseil Europénn pour la Recherche Nucléaire (European Council for Nuclear Research)
CNAO	Centro Nazionale di Adroterapia Oncologica
CRT	Conformal Radiotherapy
CT	Computed Tomography
DNA	Deoxyribonucleic Acid
EC	European Community
GSI	Gesellschaft fur Schwerionenforschung
Gy	Gray
GyE	Gray Equivalent
HIBMC	Hyogo Ion Beam Medical Center
HIMAC	Heavy Ion Medical Accelerator in Chiba
HIPLAN	Hitachi Integrated Planning
HIT	Heidelberger Ionenstrahl-Therapiezentrum (Heidelberg Ion Therapy Center)
HSG	Human Salivary Gland
IBA	Ion Beam Application
IGRT	Image Guided Radiotherapy
IMHT	Intensity Modulated Hadron Therapy
IMRP	Intensity Modulated Radiation Therapy
INFN	Istituto Nazionale di Fisica Nucleare
LBNL	Lawrence Berkeley National Laboratory
LC	Local Control
LEM	Local Effect Model
LET	Linear Energy Transfer
Linac	Linear Accelerator
MCS	Multiple Coulomb Scattering
MRI	Magnetic Resonance Imaging
NIRS	National Institute for Radiobiological Sciences
OER	Oxygen Enhancement Ratio
PET	Positron Emission Tomography

PSI	Paul Scherrer Institute
RBE	Relative Biological Efficiency
RF	Radiofrequency
SB	Skull Base
SOBP	Spread-Out Bragg Peak
TRIUMF	Tri-University Meson Facility

Chapter 1
Introduction

1.1 A Brief History of Radiotherapy and Types of Radiation

Approximately 2 months after the discovery of X-rays by Roentgen in December 1895 [1], the use of the therapeutic properties of radiation began, being used initially as a treatment for patients with or without cancer. In current times, radiation therapy is used exclusively for curative and palliative treatments, often in combination with surgery and chemotherapy. Radiation therapy can be divided into two main categories:

1. Brachytherapy, which employs sealed radioactive sources and unsealed, placed near or within the tumor
2. Teletherapy, which uses external radiation beams, through a particle accelerator or radioactive sources

Radiotherapy may be provided directly or through a secondary source of radiation, such as X-ray generation by electrons bombarding a target. Treatments may also occur in two phases, such as boron-neutron-capture therapy (BNCT) [2], where neutrons are absorbed by the core of ^{10}B, which connected (or dispersed) in tumor cells, originate alpha particle emitters and thus a local radiation dose (Fig. 1.1).

When talking about gamma radiation and X-rays, it should always be remembered that they come from different sources, with gamma radiation resulting from nuclear decay. Both consist of extremely small wavelength photons and are capable of causing ionization when going through biological environments; therefore, they are called ionizing radiation. From all the techniques used in internal or external radiotherapy, they are the most common techniques, mainly because of economic factors.

Figure 1.2 shows the different types of radiation: X-rays, gamma rays, electrons, protons, neutrons, negative pi meson, carbon ions, and neon [3]. Currently, X-rays are most commonly used due to the low price of linear accelerators (linacs).

M. d'Ávila Nunes, *Protontherapy Versus Carbon Ion Therapy*,
Biological and Medical Physics, Biomedical Engineering,
DOI 10.1007/978-3-319-18983-3_1

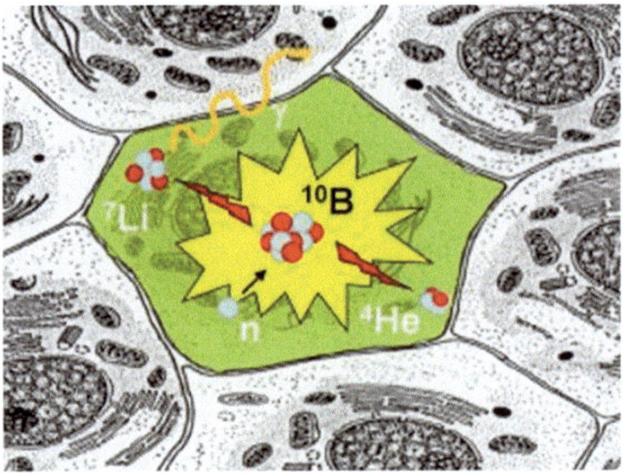

Fig. 1.1 Artistic description of BNCT. The ^{10}B atom, previously charged into the tumor cell, undergoes nuclear reaction when it absorbs a thermal neutron. The short-range, high linear energy transfer (LET) reaction fragments and destroys the tumor cell. *Courtesy* Prof. Dr. Angela Bracco, NuPECC Chair

In hadron therapy, the protons and carbons ions stand over the others. Neon has been widely used in initial research using charged particles.

Radiation may kill cancer cells, breaking the DNA molecules and preventing cell replication. X-rays can break DNA or pass through its structure; however, protons are more lethal and carbon ions are two to three times more efficient than X-rays (Fig. 1.3) [4].

As shown in Fig. 1.4, the highest density of secondary electrons is produced by carbon ions, leading to a greater break of clustered DNA [5].

The biological system has the ability to fix injuries that occur in DNA. However, if DNA is exposed to a high local dose of radiation, the repair fails to correct the damage at the most effective dose compared to ionizing radiation. Thus, the impact of radiation on the microscopic level, see Fig. 1.5 where 53BP1 protein and RPA, both related to DNA repair, are made fluorescent by immunostaining [6].

Jakob et al. [7] provided the image shown in Fig. 1.5, employing a 9.5 MeV 12C beam to irradiate a monolayer of cells, which was visualized using a microscope.

1.2 Cancer: Statistical Considerations

Cancer can be defined as the uncontrolled growth and proliferation of a group of cells. In 1982, 1.2 million new cancer cases were diagnosed in Europe. Three years later, 750,000 deaths were attributed to cancer, with death from cancer occurring in approximately 20 % of cases. In developed countries, about 30 % of the population is diagnosed with cancer, and about half die of this disease. This corresponds to about a million deaths per year. Certainly, the prognosis of individual cases varies

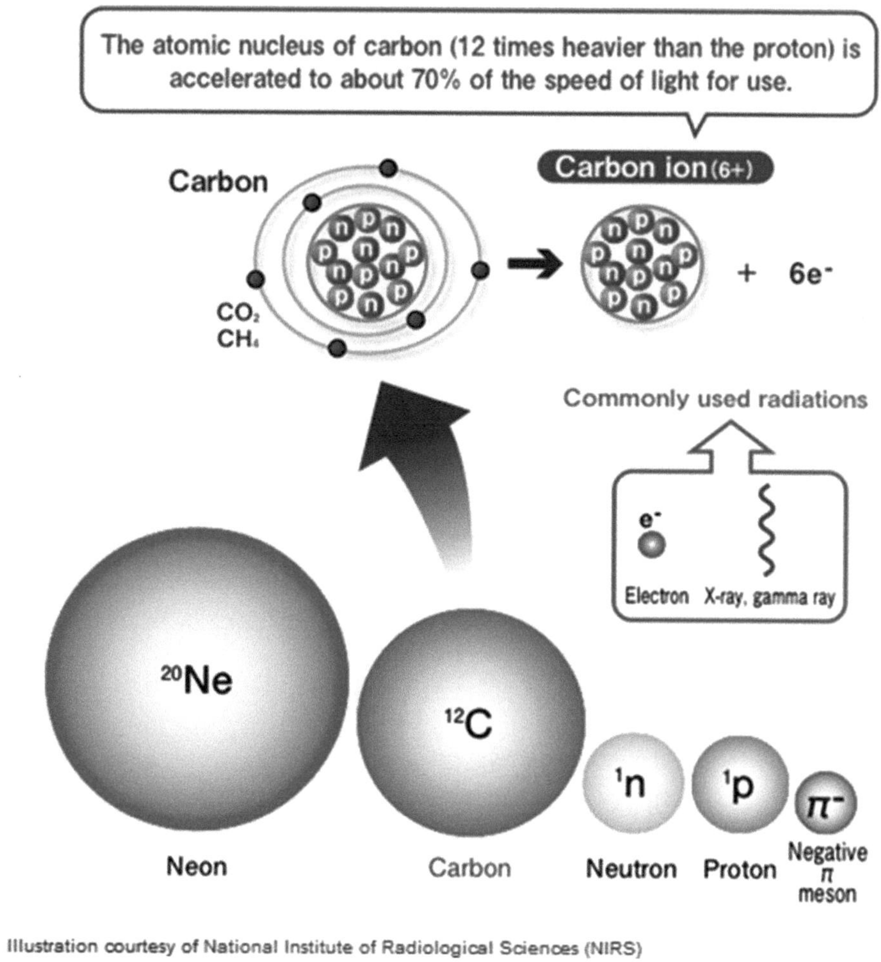

Fig. 1.2 Types of radiation used in various radiotherapy techniques. *Courtesy* National Institute of Radiological Sciences (NIRS)

and depends on the tumor type, stage, diagnosis, general health of the patient. In Europe, 45 % of patients have survived without symptoms for a period of 5 years or more.

In Russia, there are 2.3 million patients with cancer, with 450,000 new cases each year. Hadron therapy is the recommended treatment for 50,000 of these patients annually, but the capacity for this treatment by hospitals that have hadron therapy is 1,000 patients per year. Therefore, about 30–40 new protontherapy centers and 10–15 new carbon ion therapy centers should be built in Russia [8].

As a cause of death in developed countries, cancer ranks third after heart disease and stroke; it ranks second in the United States after heart disease. In 2000, studies showed that there were 10 million new cases of cases, with 6 million deaths, and

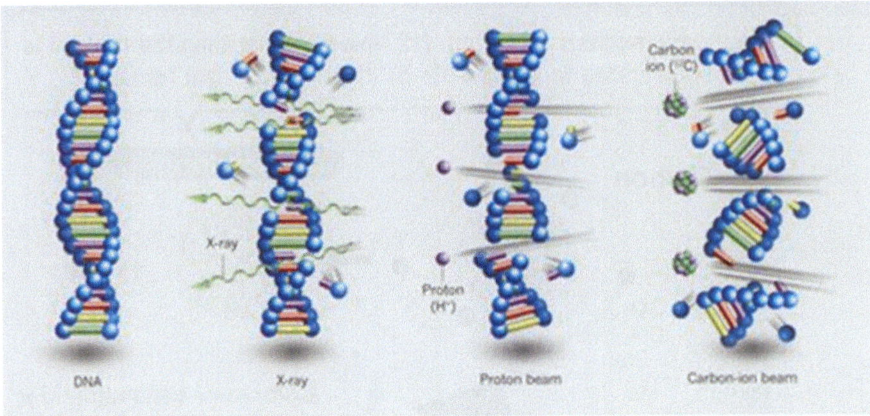

Fig. 1.3 Schematic representation of DNA breaks by type of radiation. *Courtesy* National Institute of Radiological Sciences (NIRS)

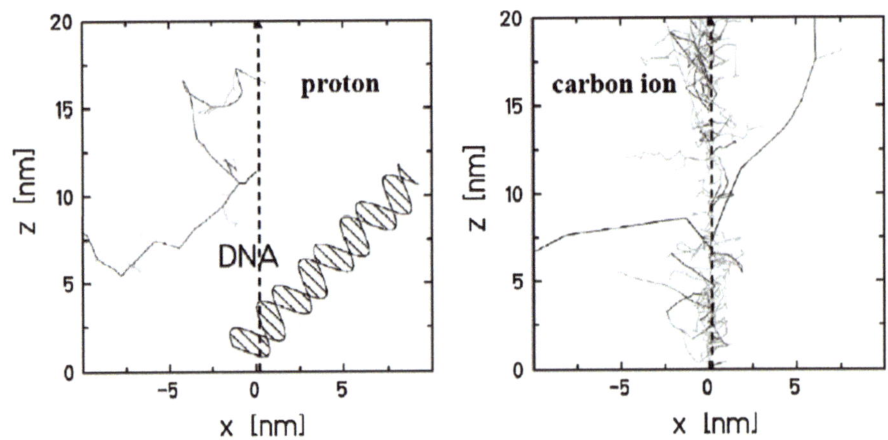

Fig. 1.4 Proton and carbon track structure in nanometric resolution compared with the schematic representation of a DNA molecule. *Courtesy* Prof. Dr. Ugo Amaldi

22 million people living with cancer worldwide [9]. These numbers represent an increase of 22 % in incidence and mortality from the year 1990 [9]. The number of new cancer cases worldwide was projected to be 12.3 to 15.4 million in 2010 and 2020, respectively [9]. In 2008, a total of 1,437,180 new cases and 565,650 cancer deaths were estimated to occur in the United States alone [9].

There are several approaches to the treatment of a malignant tumor:

(1) *Surgery* (direct removal of tissues affected by cancer): This is an invasive method and not always possible; it accounts for 22 % of treatment success.

Fig. 1.5 Repair of DNA after irradiation. 53BP1 and RPA proteins exhibit fluorescence. Reproduced from [15]

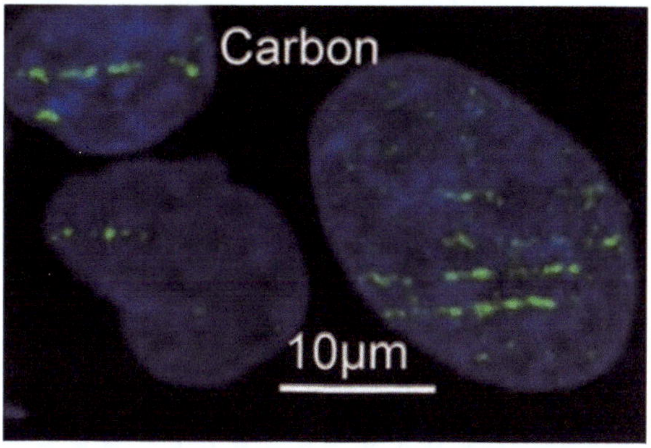

(2) *Chemotherapy* (administered drugs that prevent mitosis and cause cell death [apoptosis]). Chemotherapy causes severe side effects due to the nonspecific action of drugs in body cells.
(3) *Immunotherapy* (treatment of disease by inducing, enhancing or suppressing an immune response): Immunotherapy uses the body's own immune system to help fight cancer.
(4) *Hormone therapy* (drugs for inhibiting the activity of hormones that influence tumor growth): It is used in treatment of breast and prostate cancer, particularly with orally administration without side effects.
(5) Cell therapy, genetic treatments, and novel specific targets.
(6) *Radiation therapy* (cells are killed by energy deposition): This.has side effects due to damage to healthy tissues (in conventional radiotherapy). Radiation therapy can be administered externally by means of photons (the most widely used energy deposition method) or protons and ions (the method of tomorrow).

Of these approaches, the most important are surgery, radiotherapy, and chemotherapy. Currently, 70 % of cancer patients receive radiotherapy in the course of their treatment. Of those cured, 49 % are cured with surgery, 40 % by radiotherapy, and 11 % by chemotherapy [10]. Figure 1.6 shows the incidence and cancer mortality for all ages and both sexes [10].

Photons with high energy, which reached the megavoltage (MV) range around the year 1950, contributed significantly to the improvement of therapeutic outcomes, as shown in Table 1.1 [11].

Protons and carbon ions are the most widely used particles in the treatment of cancer worldwide. The ion beam deposits most of its energy at the end of its range, resulting in a Bragg peak (discovered by Sir William Bragg, an English physicist, in 1904). Forty-two years later, Robert R. Wilson recognized the advantage of this peak in cancer research, publishing the essential work on protons and heavy ions for the treatment of human cancer [12]. This was the first study on the application of charged particles for use in the medical field. During World War II, Wilson participated in the construction of the atomic bomb in Los Alamos; after the war,

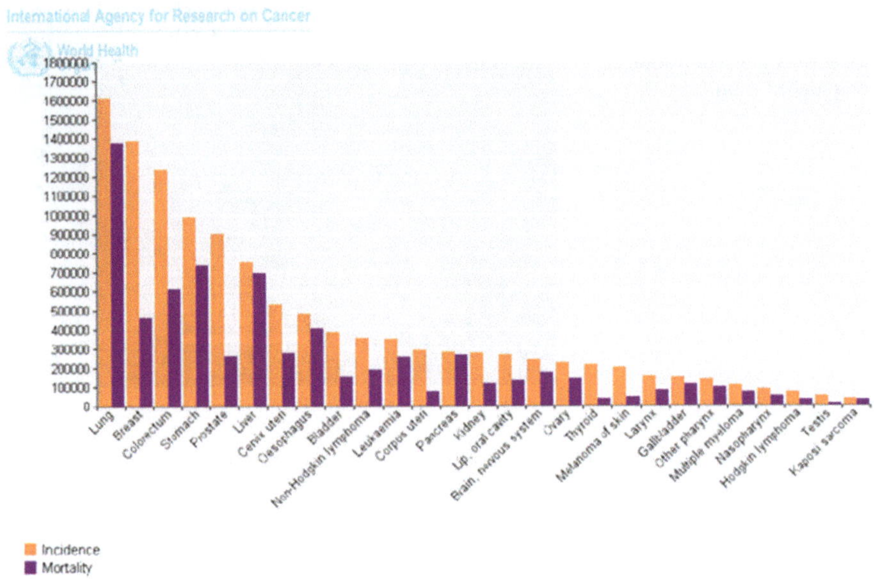

Fig. 1.6 Incidence and mortality data for all ages and both sexes. *Courtesy* International Agency for Research on Cancer, World Health Organization. Dr. Nicolas Gaudin, Head, Communications Group

Table 1.1 Improved survival of several types of cancers with the advent of megavoltage therapy

	5-year survival rate (%)	
Type of cancer	kV X-rays	MV X-rays
Hodgkin disease	30–35	70–75
Cervical cancer	35–45	55–65
Prostate cancer	5–15	55–60
Nasopharynx cancer	20–25	45–50
Bladder cancer	0–5	25–35
Ovarian cancer	15–20	50–60
Retinoblastoma	30–40	80–85
Seminoma of the testis	65–70	90–95
Embryonal cancer of the testis	20–25	55–70
Cancer of the tonsil	25–30	40–50

Courtesy From Report of the Panel of Consultants on the Conquest of Cancer. Washington, D.C., U.S. Government Printing Office, 1970. *Courtesy* Springer
These data indicated that the use of charged particles for cancer therapy may improve treatment results (Fig. 1.7)

he returned to Berkeley where he wrote a paper on the potential benefits of high-energy protons in cancer therapy. It was Wilson who proposed that carbon ions could be greater than the proton beam. He became the director of the Fermi

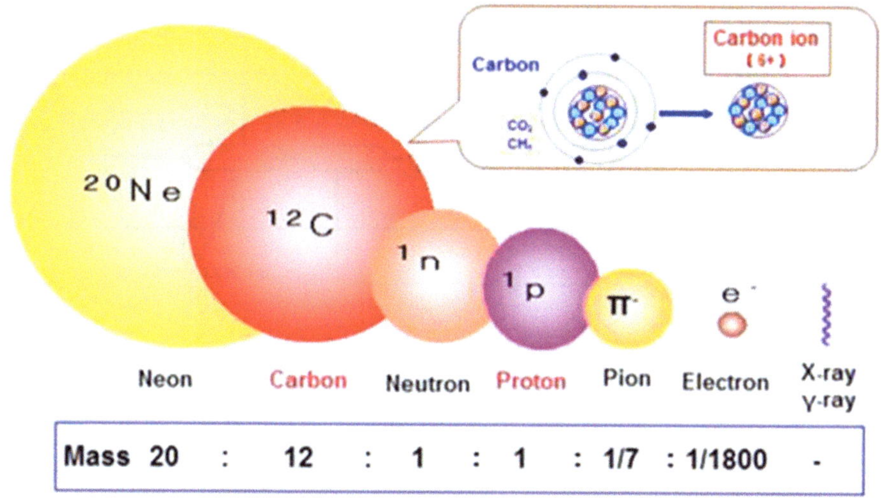

Fig. 1.7 Carbon ions have the most balanced properties of ion species in terms of both physical and biological dose distribution. *Courtesy* National Institute of Radiological Sciences (NIRS)

Laboratory, where he led the application of therapy by fast neutrons in more than 3100 patients. Compared with conventional photon therapy, particle beam therapy has minor complications and a better cure rate, and it does not affect the tissue surrounding the tumor.

1.3 Conventional Radiotherapy

The first linear accelerator (linac) was proposed in 1928 by Rolf Wideroe. In a linac (Fig. 1.8), the particles are accelerated in a straight line for a steady electric field or a field that varies with time. The best system to accelerate charged particles is to use radiofrequency (RF) fields, as high acceleration voltages can be achieved by employing RF resonant cavities compared with those obtained with similar-dimension electrostatic accelerators. Most linear accelerators proposed for hadron therapy are based on acceleration through RF fields [13].

Linear accelerators are used worldwide, treating nearly 20,000 cancer patients (for each 10 million inhabitants) in developed countries. The linacs (Fig. 1.9) replaced low-energy X-rays and gamma radiation from radioactive cobalt because they deposited the dose (energy per mass unit) at greater depths. They are extremely attractive from an economic point of view because they have a very low price compared to the circular accelerators used in hadron therapy.

As can be seen in Fig. 1.10 showing 8-MeV X-ray beams, after an initial increase of the dose absorbed, an exponential decay occurs; then, the maximum absorbed dose is reached at 2–3 cm deep in soft tissues. At a depth of 25 cm, the dose is only a third of the maximum dose.

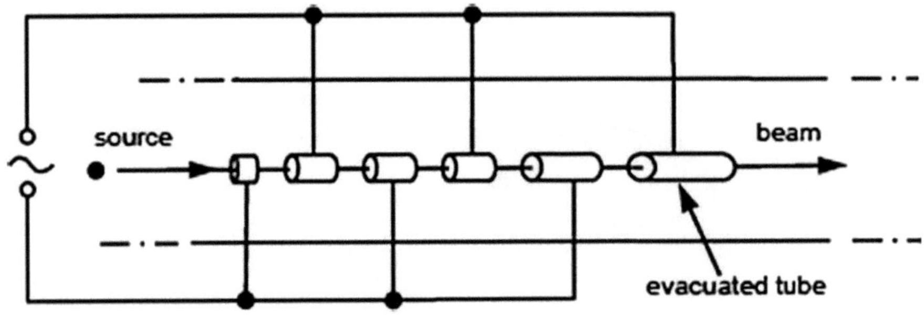

Fig. 1.8 Simplified diagram of a linear accelerator

Fig. 1.9 Conventional radiotherapy: linear accelerators dominate. *Courtesy* Prof. Dr. Ugo Amaldi

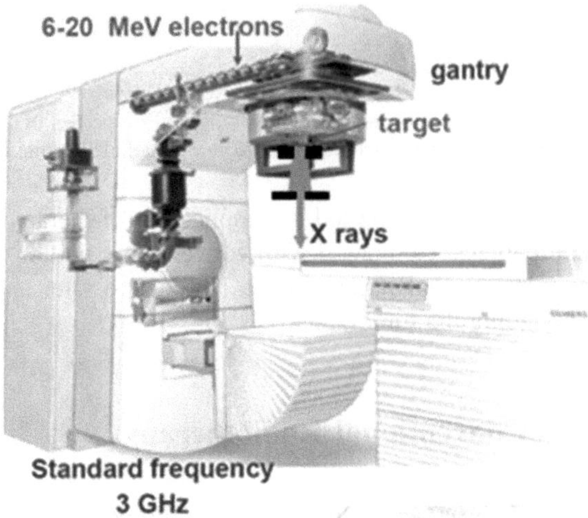

Fig. 1.10 Qualitative depth dependence of the deposited dose for each radiation type, with a narrow Bragg peak at the end. *Courtesy* Prof. Dr. Ugo Amaldi and CERN

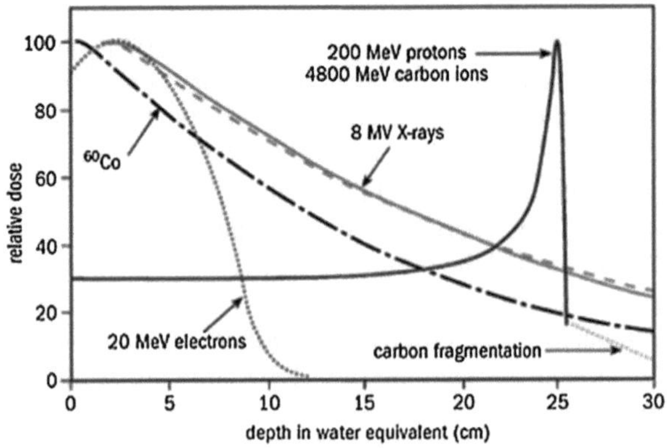

To increase the dose to the tumor, it is necessary to adjust the dose to the target. To irradiate deep tumors, multiple beams are directed to the center of the tumor (crossfire technique). A gantry is used for positioning the beam in the tumor. The technique of intensity modulated radio therapy (IMRT) uses 6–10 input ports; the beams may not be coplanar and their intensities vary via the irradiation field with the use of variable multileaf collimators, controlled by computers.

It was always necessary a high precision and greater biological effectiveness of the applied dose. High accuracy is achieved with increasing photon energy, leading to a shallow decay of the dose with a minor lateral spread. High effectiveness can be obtained by the application of hyperbaric oxygen, drugs, and heat as radiation-sensitizer agents. The high accuracy combined with the high control rate of tumor in conventional radiotherapy using X-rays has brought important results for the high energy employed in hadron therapy.

1.4 Status of Protontherapy

The history of protontherapy can be divided into two periods. In the first period, Ernest Orlando Lawrence created a system called Cyclotron in 1929, which accelerated particles to high energies without the use of high voltage, [14]. The first cyclotron was about 15 cm and fit in the palm of a hand, as shown in Fig. 1.11.

Fig. 1.11 Ernest Orlando Lawrence (1901–1958), with the first cyclotron that he built shown in the palm of his hand

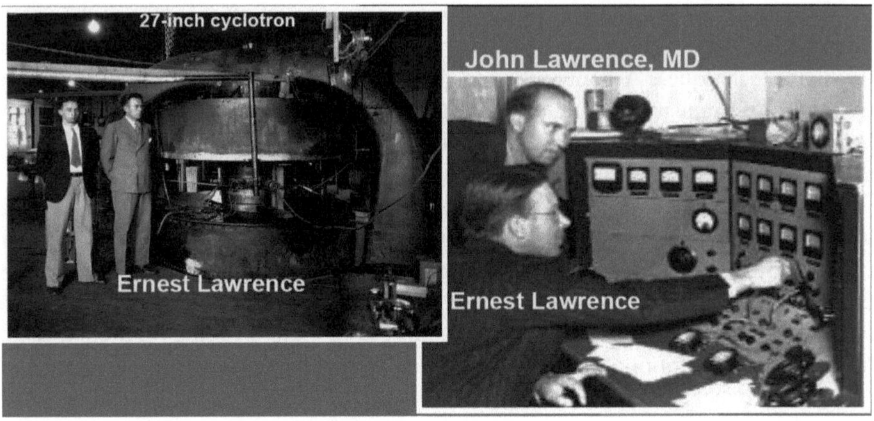

Fig. 1.12 Ernest Lawrence standing beside his 27-in cyclotron (*left*). On the *right*, the brothers Ernest Lawrence and John H. Lawrence are shown in *front* of the control panel. *Courtesy* Prof. Dr. Ugo Amaldi

Berkeley cyclotrons of 11, 27, and 37 in diameter were then built (Fig. 1.12). The latter cyclotron was used successfully in 1938 to treat 24 patients with fast neutrons. From then until 1943, a total of 226 patients were treated by fast neutrons using a 60-in cyclotron 60. However, due to side effects in healthy tissues, Stone emphasized that the technique should not be used for cancer therapy [15].

Also in 1938, a physicist at the University of Illinois, Gerald Kruger, suggested the possibility of cancer treatment using alpha particles emitted by boron when neutrons were captured. The idea was to saturate the tumor with boron and expose the tumor to a beam of neutrons. Then, the therapy technique of neutron capture emerged, since the cross-section for thermal neutron capture was approximately 100 times greater than for other tissue compounds.

In 1965, a British group re-evaluated fast neutron therapy and restarted this therapy 4 years later. It was shown that, for certain tumors, the therapy worked. Similarly, the neutron therapy resurfaced in Europe, the United States, and Japan around 1970. This therapy was gradually abandoned by institutions and used only in some hospitals for the treatment of selected tumors. There is a unit in working conditions at Argentina [15].

Protons and heavy ions have a greater mass than electrons and require large accelerators to produce sufficient kinetic energy to treat deep-seated tumors. Therefore, with the overall construction of the synchrocyclotron in 1947 at University of California by Lawrence, the acceleration of protons and heavy ions became feasible. Therefore, Lawrence suggested that his brother (physician) John H. Lawrence and the renowned Cornelius A. Tobias, work together on their 184-in cyclotron to test the ideas of Robert Wilson (Figs. 1.13 and 1.14).

The work with the 184-in cyclotron started by using a deuteron beam in the pituitary gland of female dogs with breast cancer in their experiments [16]. They were successful in their work, with tumor remission lasting for several months, but

Fig. 1.13 The 184-in cyclotron shown with the team staff (*top*). Cornelius A. Tobias (*middle*). Mounting scheme for irradiation of the pituitary gland with the included reference (*bottom*). *Courtesy* Prof. Dr. Ugo Amaldi

Fig. 1.14 This photograph of Robert Wilson was taken during the 1996 Second International Symposium on Hadron Therapy, which was held at CERN 50 years after the publication of his well-known paper, "Radiological Use of Fast Protons." *Courtesy* Prof. Dr. Ugo Amaldi and CERN

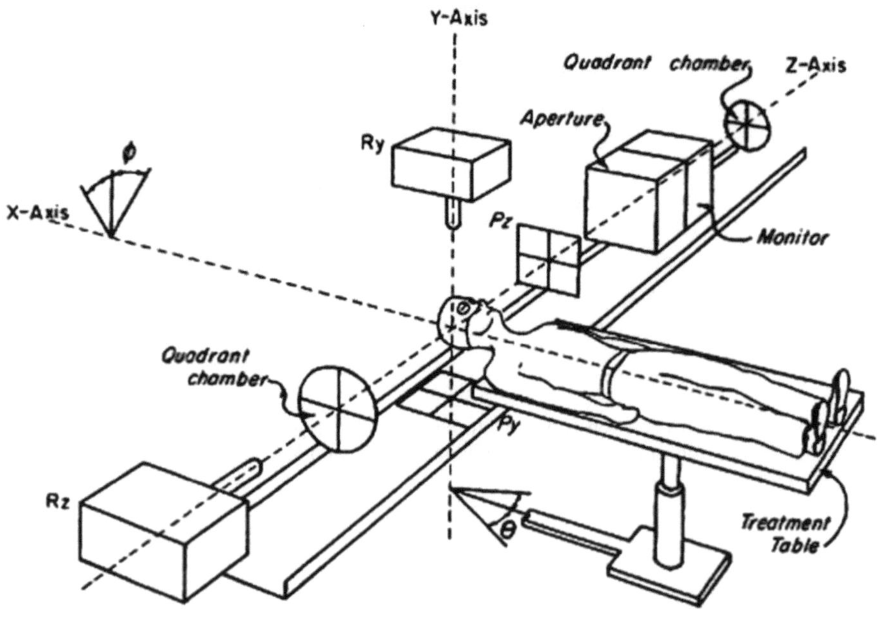

Fig. 1.15 Proton single beam setup to irradiate the pituitary (detail of Fig. 1.13). *Courtesy* Prof. Dr. Ugo Amaldi

the animals then died due to tumor relapse. The researchers decided to apply this technique in human patients with disseminated breast cancer, performing irradiation of the pituitary with a proton beam (Fig. 1.15). A noticeable effect was produced with the bleeding stopping; however, because the patients were irradiated with almost half the dose currently accepted, they died several months later. Of the 50 patients irradiated with protons, deuteron, and helium ions, half of the patients exhibited beneficial effects and others did not show any benefit. Regarding toxicity, the effect observed was the development of diplopia.

Tobias and Lawrence wrote in 1958 [16],

> The demonstration that hypophysectomy might be of benefit in the palliative treatment of various diseases such as advanced breast cancer, fulminating juvenile diabetes with retinopathy, malignant exophthalmos, and malignant hypertension prompted the inquiry into the possible use of the proton beam to destroy or inhibit the function of the pituitary gland. The technic (sic) was perfected by extensive animal investigation, in which both 190 meV deuterons and 340 meV protons were employed, but only the 340 meV proton beam was used in attempts to destroy the human pituitary, first in patients with advanced metastatic breast cancer.

Some of the more recent applications of protontherapy in the treatment of human cancer include pediatric treatment and treatment of uveal melanoma. In recent decades, institutions such as the Harvard Cyclotron Laboratory has been able to treat thousands of patients successfully and safely for these and other conditions.

The two researchers thought that protons and helium ions scattered within the head due to the multiple scattering effect, and that this effect could be minimized if oxygen or carbon beams were used.

Next, they started to radiate the pituitary gland in patients with acromegaly. They treated 700 patients and observed that the acromegaly remained in regression, with the growth hormones staying within normal limits for many years [16]. In this way, in 1957, the synchrocyclotron was modified to accelerate helium nuclei. More than 2000 patients were irradiated with helium ions. All treatment in the 1950s and 1970s was focused on pituitary tumors because the location was easy to access and allowed complete immobilization of the skull. There were still no resources using the CT scan, which only emerged in 1973.

In the second historical period of protontherapy, several protontherapy centers began to emerge in nuclear physics research facilities, including Uppsala, Sweden (1957), Cambridge, Mass. (1961), Duhna (1967), Moscow, Russia (1969), St. Petersburg, Russia (1975) Chiba, Japan (1979), Tsukuba, Japan (1983), and Villigen, Switzerland (1984) [15]. Indeed, the advent of computed tomography, which allowed the precise determination of the path of the beam in the patient, made particle therapy practical and easy. In the 1970s and 1980s, tumors of the skull base and intracranial tumors were treated, as seen in Table 1.2.

Most patients treated by protontherapy at this time had choroidal melanomas, which were treated using doses of 60–70 Gy in 4–5 divided doses per week. Tsujii et al. [11] of Tsukuba University treated deep-seated tumors in the lung, esophagus, liver, uterine cervix, prostate, head, and neck.

Table 1.2 Distribution of tumors treated with proton beams in the early phase of its clinical application in the world (as of May 1993)

Tumor sites	USA	Europe	Russia	Japan	Total (%)
Ocular melanoma	1,698	2,196	355	44	4,293 (35.1)
Skull base and upper spine	3,132	15	1,678	58	4,883 (39.9)
Head and neck	79	20	0	21	120 (1.0)
Thoraco-abdominal	2	0	0	127	129 (1.1)
Pelvis (prostate, uterus,etc.)	469	41	242	61	813 (6.6)
Others	18	12	77	128	235 (1.9)
Unknowns	709	27	1,025	0	1,761 (14.4)
Total (%)	6,107 (49.9)	2,311 (18.9)	3,377 (27.6)	439 (3.6)	12,234 (100.0)

Courtesy Springer [11]

In 1990, the Loma Linda University Medical Center emerged as the first to employ a synchrotron dedicated to medical care and research. They had 4 rooms, 3 gantries, and a fixed beam line. The center used technicians working at Fermilab with the excellent physicist Don Lincoln for the construction of a low-energy synchrotron (up to 250 MeV). After this point, protontherapy units around the world emerged [15].

Figure 1.16 shows the number of patients who have been treated with protontherapy worldwide as of December 2013.

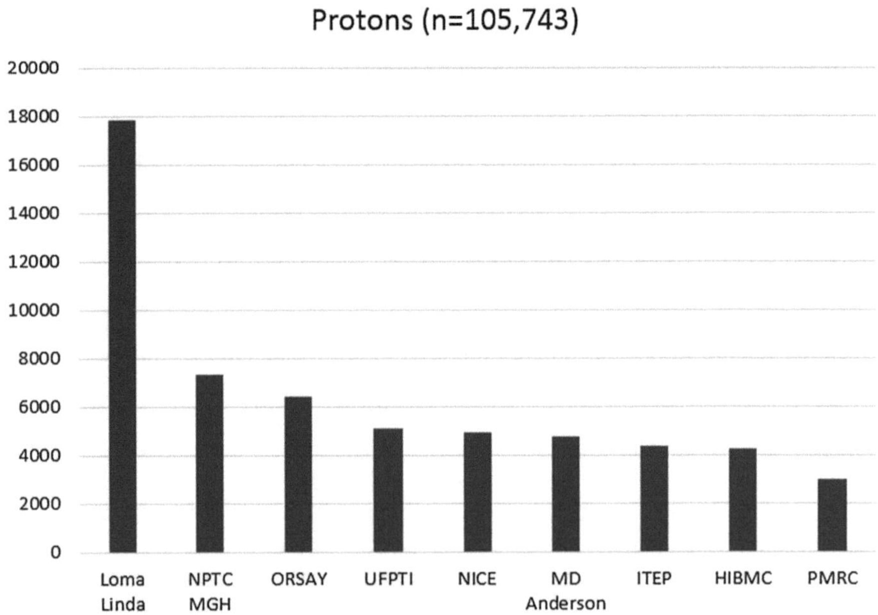

Fig. 1.16 Number of patients treated at proton facilities worldwide, according to the PTCOG web site (as of December 2013). *Source* http://ptcog.ch

1.5 Status of Carbon-Ion Therapy

The construction of the Bevatron (based on synchrotron) in Berkeley in 1954 opened the possibility of obtaining carbon-ion beams. Certainly, it could also be used for research with oxygen and neon particles. The Bevatron contributed for more than 40 years in various areas of research. It was replaced 20 years later by the Bevalac, which is a Bevatron combined with a linear accelerator (SuperHILAC). Therefore, in Berkeley began the therapy for helium ions and 20 years later for neon-ion therapy.

In 1977, the first treatment by carbon ions was performed at Lawrence Berkeley National Laboratory (LBNL) [11]. Berkeley ended its activities after 17 years, having treated more than 2,000 patients. However, these studies encouraged other centers around the world to use carbon ion therapy, such as the National Institute of Radiological Sciences (NIRS) in Japan, through its Heavy Ion Medical Accelerator in Chiba (HIMAC, Fig. 1.17). Thereafter, carbon ion therapy was initiated in Japan in 1994, with clinical applications [11].

The experience at Berkeley with carbon ions was transferred to Chiba by Japanese researchers who used to be interns there. The Germans, recognizing that the technique was promising, set up a center in Gesellschaft fur Schwerionenforschung (GSI) in Germany in 1997 [15], which was succeeded by the Heidelberg Ion-Beam Therapy Center (HIT) in 2000. The HIT works with protons and carbon ions; they also developed the raster scan method, which allows the ion beam to go inside the tumor pixel by pixel and then deposit the required amount of energy for the optimal breakdown of the tumor DNA [15]. HIT is a leading center in the world today for carbon ion therapy.

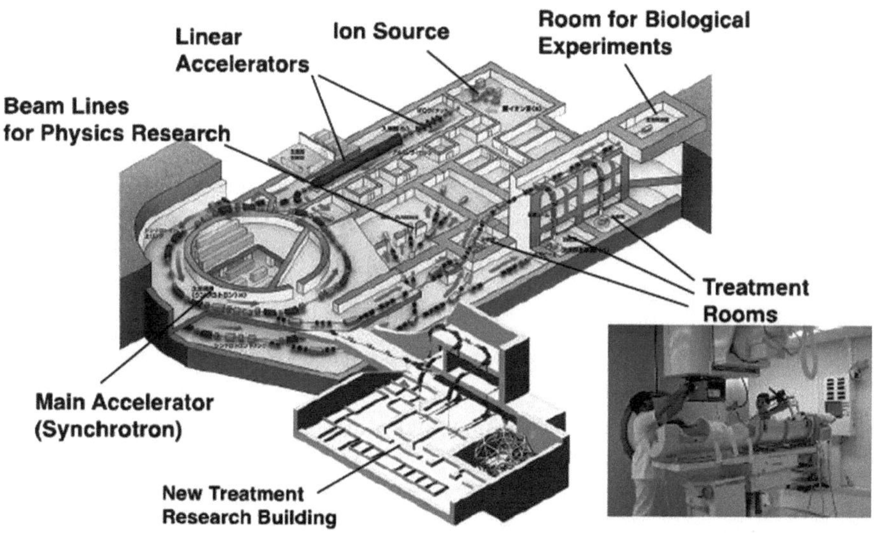

Fig. 1.17 Overview of HIMAC in Chiba, the world's first accelerator complex dedicated to cancer therapy. It was built in 1993 at NIRS. *Courtesy* Springer [11]

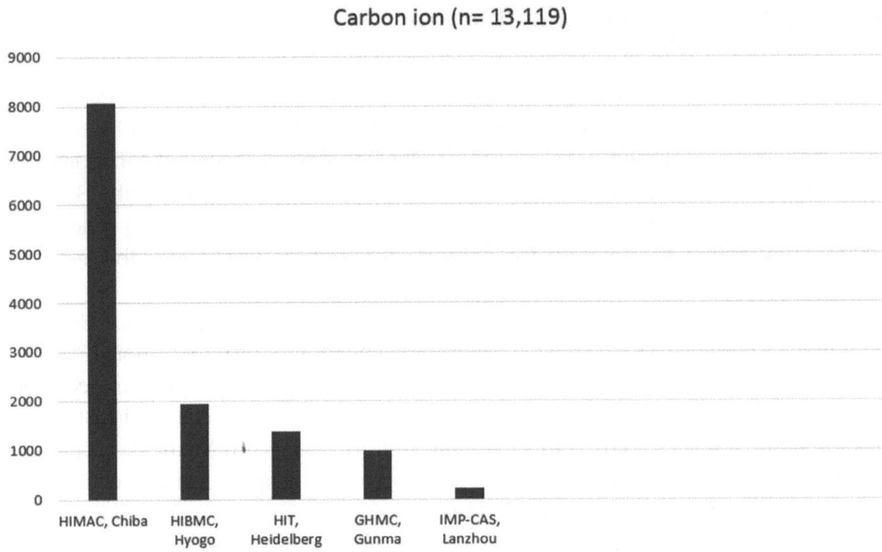

Fig. 1.18 Number of patients treated at carbon-ion facilities in the world, according to the PTCOG web site (as of December 2013). *Source* http://ptcog.ch tables

Figure 1.18 shows the number of patients treated by carbon ion therapy worldwide.

There are six carbon-ion therapy centers in operation—three in Japan (NIRS; the Hyogo Ion Beam Medical Center in Hyogo, Japan, 2002; and Gunma University Heavy Ion Medical Center in Gunma, Japan, 2010), one in Germany (HIT), one in Italy (Centro Nazionale di Adroterapia Oncologica in Pavia, Italy, 2011) and one in China (Institute of Modern Physics in Lanzhou, 2006). Several facilities are in the construction phase—two facilities in Japan, one in Germany, one in Austria, and two in China. The United States only has protontherapy centers; however, they already have funds to start construction at the University of Colorado and a center in Michigan supported by a consortium. The interest in carbon ion therapy has increased in several US hospitals that are associated with the Japanese researchers NIRS, who are creating symposia on carbon ion therapy, such as the Department of Radiation Oncology at the Mayo Clinic (Joint Symposium on Carbon 2013 ion Radiotherapy).

Currently, 38 hadron therapy centers are in operation worldwide :

- Europe: 11 centers distributed in Italy, France, Germany, England, Switzerland, Sweden, Poland, Russia
- Asia: 8 centers in Japan, 2 in China, and 1 in South Korea
- America: 11 centers in the USA and one in Canada
- South Africa : 1 center

Figures 1.19, 1.20 and 1.21 show in detail the location of each center. Table 1.3 shows further details on the centers, as updated and organized by PTCOG. PTCOG

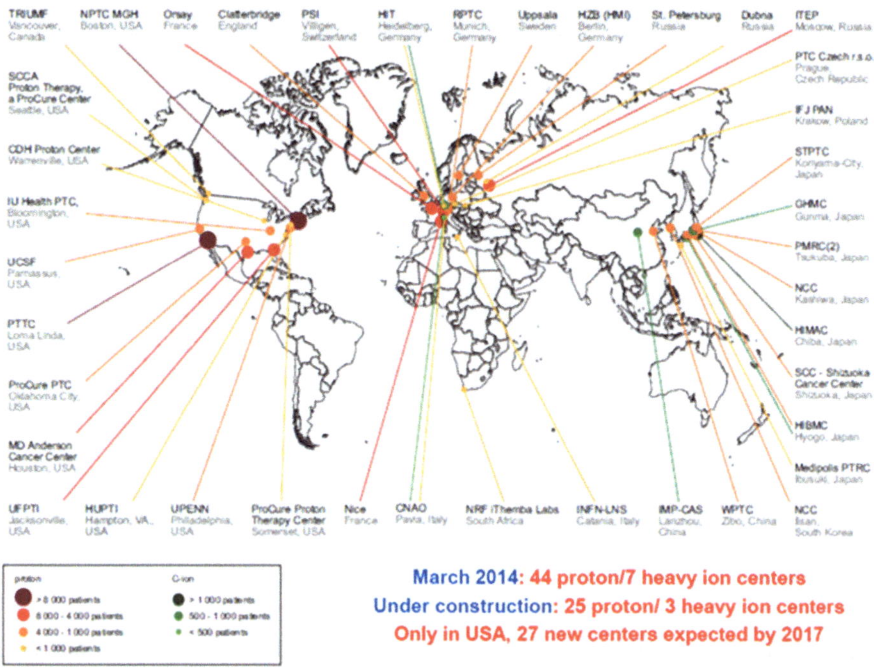

Fig. 1.19 Proton (*red-orange*) and C-ion (*green*) centers active worldwide. The size of the spot is proportional to the number of patients treated, as indicated in the figure legend. Note that there are no centers in South and Central America. In Central America, there is a protontherapy center under construction (equipment from IBA). *Courtesy* Prof. Dr. Angela Bracco, NuPECC Chair

is an organization that provides important data on radiation therapy with protons, light ions and heavy particles to benefit patients in need of hadron therapy (http://ptcog.ch).

1.6 Hadron Therapy Timeline

This section examines some historical data on hadron therapy and their timeline (Fig. 1.22) [15].

1. The first cancer treatment of deep-seated tumors using radiation (X-rays) was carried out by the brothers Lawrence (Ernest O. Lawrence and John Lawrence) in 1937. The treatment seemed to have cured the uterine cancer (inoperable) of their mother, but the disease was probably misdiagnosed.
2. JS Stone and John Lawrence, both medical doctors, used neutron therapy in patients starting in 1938, with a program that involved 250 patients. Stone concluded that neutron therapy was a "delayed stressful" modality and "should

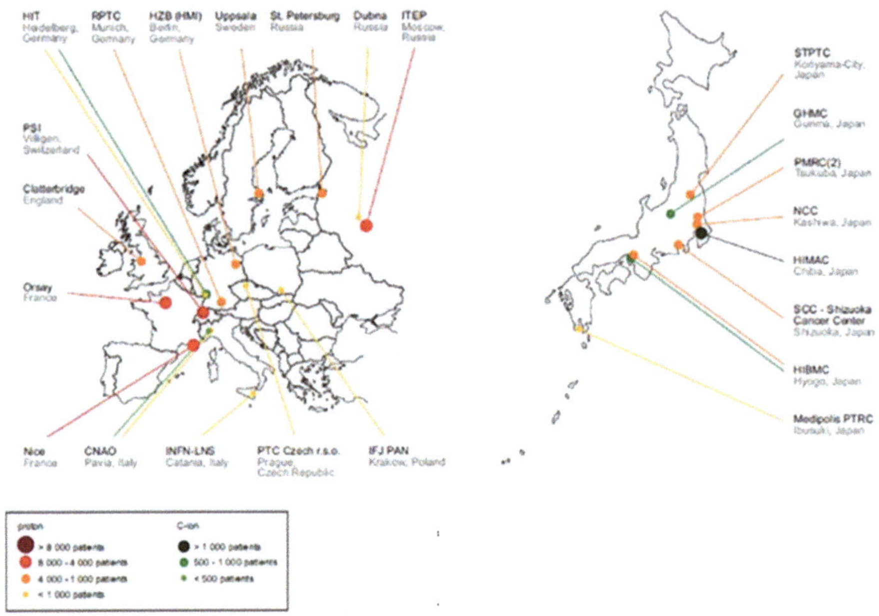

Fig. 1.20 Proton (*red-orange*) centers active in Europe and Japan. The size of the spot is proportional to the number of patients treated, as indicated in figure legend. *Courtesy* Prof. Dr. Angela Bracco, NuPECC Chair

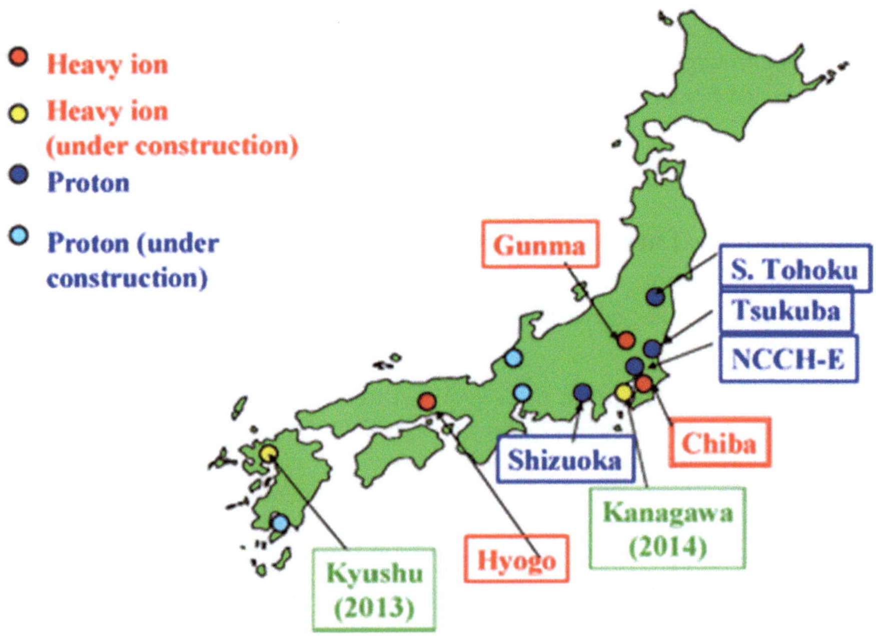

Fig. 1.21 Heavy-ion and proton radiotherapy facilities in Japan. *Courtesy* National Institute of Radiological Sciences (NIRS)

Table 1.3 Hadron therapy facilities worldwide

PTCOG Secretary, June 2014

Particle Therapy Patient Statistics (per end of 2013)

WHERE COUNTRY	SITE	PARTICLE	FIRST (-LAST) PATIENT	PATIENT TOTAL	DATE OF TOTAL	
Belgium	Louvain-la-Neuve	p	1991 (-1993)	21	1993	ocular tumors only
Canada	Vancouver (TRIUMF)	π^-	1979 (-1994)	367	1994	ocular tumors only
Canada	Vancouver (TRIUMF)	p	1995	175	Dec-13	ocular tumors only
Czech Rep.	Prag (PTCCZ)	p	2012	140	Dec-13	
China	Wanjie (WPTC)	p	2004	1078	Dec-13	
China	Lanzhou	C Ion	2006	213	Dec-13	
England	Clatterbridge	p	1989	2446	Dec-13	ocular tumors only
France	Nice (CAL)	p	1991	4936	Dec-13	ocular tumors only
France	Orsay (CPO)	p	1991	6432	Dec-13	5082 ocular tumors
Germany	Darmstadt (GSI)	C-Ion	1997 (-2009)	440	2009	
Germany	Berlin (HMI)	p	1998	2312	Dec-13	ocular tumors only
Germany	Munich (RPTC)	p	2009	1811	Dec-13	
Germany	HIT, Heidelberg	C Ion	2009	1368	Dec-13	
Germany	HIT, Heidelberg	p	2009	503	Dec-13	
Germany	WPE, Essen	p	2013	32	Dec-13	
Italy	Catania (INFN-LNS)	p	2002	293	Nov-12	ocular tumors only
Italy	Pavia (CNAO)	p	2011	76	Dec-13	
Italy	Pavia (CNAO)	C Ion	2012	105	Dec-13	
Japan	Chiba	p	1979 (-2002)	145	2002	ocular tumors only
Japan	Tsukuba (PMRC, 1)	p	1983 (-2000)	700	2000	
Japan	Chiba (HIMAC)	C Ion	1994	8073	Dec-13	377 with scanning
Japan	Kashiwa (NCC)	p	1998	1226	Mar-13	
Japan	Hyogo (HIBMC)	p	2001	4223	Dec-13	
Japan	Hyogo (HIBMC)	C Ion	2002	1935	Dec-13	
Japan	WERC	p	2002 (-2009)	62	2009	
Japan	Tsukuba (PMRC, 2)	p	2001	2967	Dec-13	
Japan	Shizuoka	p	2003	1590	Dec-13	
Japan	Koriyama-City	p	2008	2306	Dec-13	
Japan	Gunma	C Ion	2010	985	Dec-13	
Japan	Ibusuki (MMRI)	p	2011	919	Dec-13	
Japan	Fukui City (Prefectural Hospital)	p	2011	428	Dec-13	
Japan	Nagoya PTC, Nagoya, Aichi	p	2013	199	Dec-13	
Japan	Tosu (Saga-HIMAT)	p	2013	62	Dec-13	
Poland	Krakow	p	2011	39	Dec-13	ocular tumors only
Russia	Dubna (1)	p	1967 (-1996)	124	1996	
Russia	Moscow (ITEP)	p	1969	4320	Dec-13	
Russia	St. Petersburg	p	1975	1386	Dec-12	
Russia	Dubna (JINR, 2)	p	1999	995	Dec-13	
South Africa	iThemba LABS	p	1993	521	Dec-13	
South Korea	Ilsan, Seoul (NCCR)	p	2007	1266	Dec-13	
Sweden	Uppsala (1)	p	1957 (-1976)	73	1976	
Sweden	Uppsala (2)	p	1989	1356	Dec-13	
Switzerland	Villigen PSI (Piotron)	π^-	1980 (-1993)	503	1993	
Switzerland	Villigen PSI (OPTIS 1)	p	1984 (-2010)	5458	2010	ocular tumors only
Switzerland	Villigen-PSI, incl OPTIS2	p	1996	1581	Dec-13	695 ocular tumors
USA, CA.	Berkeley 184	p	1954 (-1957)	30	1957	
USA, CA.	Berkeley	He	1957 (-1992)	2054	1992	
USA, NM.	Los Alamos	π^-	1974 (-1982)	230	1982	
USA, CA.	Berkeley	ions	1975 (-1992)	433	1992	
USA, MA.	Harvard (HCL)	p	1961 (-2002)	9116	2002	
USA, CA.	Loma Linda (LLUMC)	p	1990	17829	Dec-13	
IN., USA	Bloomington (MPRI, 1)	p	1993 (-1999)	34	1999	ocular tumors only
USA, CA.	UCSF - CNL	p	1994	1621	Dec-13	ocular tumors only
USA, MA.	Boston (NPTC)	p	2001	7345	Dec-13	
USA, IN.	Bloomington (IU Health PTC)	p	2004	1927	Dec-13	
USA, TX.	Houston (MD Anderson)	p	2006	4746	Dec-13	
USA, FL	Jacksonville (UFPTI)	p	2006	5085	Dec-13	
USA, OK.	Oklahoma City (ProCure PTC)	p	2009	1364	Dec-13	
USA, PA.	Philadelphia (UPenn)	p	2010	1750	Dec-13	
USA, IL.	CDH Warrenville	p	2010	1329	Dec-13	
USA, VA.	Hampton (HUPTI)	p	2010	767	Dec-13	
USA, NY.	New Jersey (ProCure PTC)	p	2012	512	Dec-13	
USA, WA	Seattle (SCCA ProCure PTC)	p	2013	86	Dec-13	
USA, MO.	St. Louis (S. Lee King PTC)	p	2013	1	Dec-13	

Total for all facilities (in operation and out of operation):

2054	He
1100	pions
13119	C-ions
433	other ions
105743	protons
122449	**Grand Total**

Last update: June 2014. *Courtesy* PTCOG

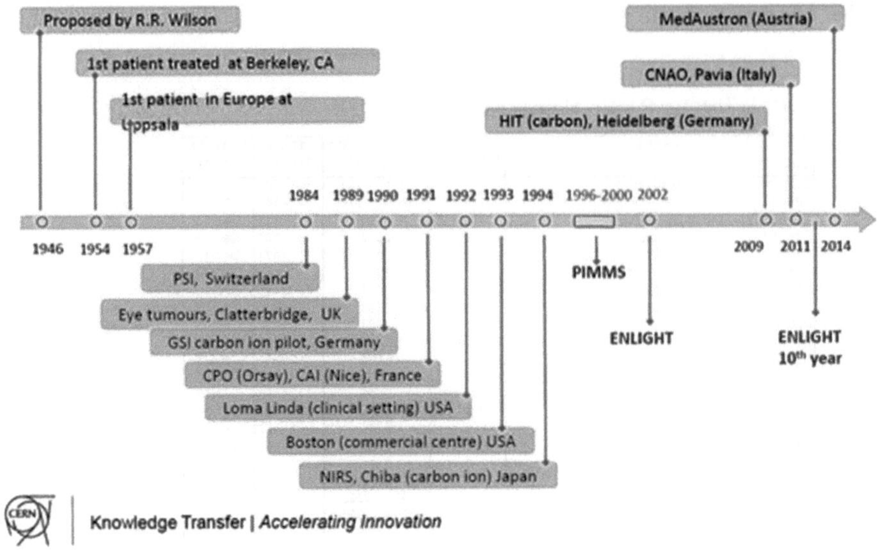

Fig. 1.22 Proton and ion beam therapy: a short history. *Courtesy* CERN Courier

not be continued". No additional work was undertaken regarding neutron therapy over the following 25 years.

3. Siemens and Varian of the USA built the first X-ray linacs.
4. Most patients are treated using X-rays. There are 10,000 linacs worldwide and they are used to treat 4,000,000 patients per year.
5. Hadron therapy (Bragg peak) was first suggested by Robert Wilson [31].
6. Berkeley and Harvard Universities were pioneers in this area.
7. The combination of the Bevatron particle accelerator with the SuperHILAC (linear accelerator used as an injector for heavy ions) was named the Bevalac. During the 1970s, the use of heavy ions was carefully developed in the Bevalac (the only accelerator capable of accelerating any nucleus in the periodic table to relativistic energies) from basic biology to patient treatment. Great effort was expended on research and development to answer questions involving which types of cancer responded best to treatment and the optimal radiation doses. Many scientists, including Joe Castro, Bill Chu, John Lyman, Cornelius Tobias, Eleanor Blakely, Ted Philips and others, participated in these studies. The Bevalac was used two thirds of the time for medical studies and one third of the time for nuclear physics studies. The groundwork was laid for the relativistic heavy ion collider (RHIC) and the large hadron collider (LHC).
8. Based on the work carried out at Berkeley, the heavy ion medical accelerator in Chiba (HIMAC) was built in Chiba, Japan. It was the first facility dedicated to the treatment of cancer using ions. Although none of these accelerators were established in the United States, many were built in Japan (approximately 50) and some were also built in Europe.

9. Therapy using ions and neutrons has been used in the past, but it did not prove to be of great interest to the oncology community. However, treatment using fast neutrons was initiated at Fermilab in the United States.

History of the medical linear accelerator:

1952: Henry Kaplan and Edward Ginzton begin building a medical linear accelerator.
1956: The first medical linear accelerator in the Western Hemisphere is installed at Stanford Hospital in San Francisco.
1959: Stanford Medical School and Hospital move to the Palo Alto campus, bringing the medical linear accelerator.
1962: Kaplan and Saul Rosenberg begin trials using the linear accelerator with chemotherapy to treat Hodgkin's disease, an approach that dramatically improves patient survival.
1994: First use of the CyberKnife, invented at Stanford, which uses sophisticated computerized imaging to aim a narrow X-ray beam precisely.
1997: Stanford pioneers the use of intensity-modulated radiation therapy, which combines precise imaging with linear accelerators that deliver hundreds of thin beams of radiation from any angle.
2004: Implementation of four-dimensional radiotherapy, which accounts for the motion of breathing during imaging and radiation delivery.
2006: 50th anniversary of the Stanford medical linear accelerator.

1.6.1 Timeline Key Event (in Detail)

1930: Ernest Lawrence invents the cyclotron accelerator.
1938: Neutron therapy is developed by John Lawrence and JS Stone at Berkeley University.
1946: For the first time, Robert Wilson suggests that energetic protons could be an effective cancer treatment method in a study published in 1946, while he was involved in designing the Harvard Cyclotron Laboratory.
1948: Extensive studies in Berkeley confirm Wilson's suggestion.
1954: Protons are used to treat patients in Berkeley.
1957: Uppsala University, Sweden duplicates the results obtained at Berkeley.
1961: The Harvard Cyclotron Laboratory teams up with Massachusetts General Hospital to use protontherapy. The first treatment takes place at Harvard. Over a period lasting until 2002, a total of 9,111 patients were treated. The hospital was closed in 2002.
1968: Installation of a particle accelerator takes place in Dubna, Russia.
1969: Installation of a proton accelerator in Moscow is completed.
1970: The Massachusetts General Hospital conducts the first study regarding radiotherapy with protons/photons for the treatment of prostate cancer.

1972: Fast neutron therapy is initiated at the MD Anderson Hospital in Texas (soon more units open at six locations in the United States).
1974: Patients are treated with a pi-meson beam at the Los Alamos National Laboratory in New Mexico. Treatment is terminated at the end of 1981; it was started and finished at the Paul Scherrer Institute (PSI) in Switzerland and the Tri-University Meson Facility (TRIUMF) facility in Canada.
1975: A protontherapy facility is opened in St. Petersburg, Russia.
1975: A team at Harvard University pioneer the treatment of eye cancer with protons.
1976: Fast neutron therapy is started at Fermilab. This facility is closed in 2003 after treating 3,100 patients.
1977: Cancer treatment with ions is initiated using the Bevalac. The facility is closed in 1992, after treating 233 patients.
1979: Protontherapy commences in Chiba, Japan, using the HIMAC.
1980: The design and construction of the first facility dedicated to clinical protontherapy treatments takes place at Loma Linda University Medical Center in California.
1988: Protontherapy is approved by the Food and Drug Administration (FDA) in the United States.
1989: Protontherapy commences using the Clatterbridge accelerator in the United Kingdom.
1990: Medical plans covering protontherapy are developed by the Particle Therapy Co-Operative Group (PTCOG): www.ptcog.web.psi.ch
1990: The first hospital-based proton treatment facility is opened in Loma Linda, California.
1991: Protontherapy is initiated in Nice and Orsay in France.
1992: The Berkeley cyclotron accelerator is closed after treating more than 2,500 patients.
1993: Protontherapy is initiated in Cape Town, South Africa.
1993: Indiana University Health Protontherapy Center in the United States treats its first patient with protons.
1994: Therapy with carbon ions is started at the HIMAC in Japan. By 2008, more than 3,000 patients have been treated.
1996: Installation of a proton accelerator at the PSI in Switzerland.
1998: Installation of a proton accelerator in Berlin, Germany.
1990–2000: More than 25,000 patients are treated worldwide with protontherapy.
2001: Massachusetts General Hospital in the United States opens a protontherapy center.
2006: The MD Anderson Hospital in Texas opens a protontherapy center.
2007: A protontherapy center is opened in Unity, Jacksonville, Florida.
2008: Neutron therapy is restarted at Fermilab.

Only the United States, Europe, Asia, and Africa have hadron treatment resources. There are no hadron therapy facilities in South America; there is only one ongoing project in Argentina that involves a related technology known as boron neutron capture therapy (BNCT). Locations of hadron therapy facilities around the world are detailed below; the latest updates are available at the website (ptcog.web.psi.ch).

- In the United States:
  - Loma Linda, CA (1990); Boston, MA (2001); Bloomington, IN (2004): Houston, TX (2006); Jacksonville, FL (2006). (Ion therapy formally took place at the Berkeley Laboratory, CA)
- In the rest of the world:
  - Japan: Chiba (1994); Kashiwa (1998); Tsukuba (2001); Hyogo (2001); Wakasa (2002): Shizuoka (2003), Tsunuga.
  - Germany: Munich; Essen; Heidelberg (HIT); Marburg; Kiel.
  - Europe: Pavia, Italy; Orsay, France; Trento, Italy; Uppsala, Sweden; Vienna, Austria; Lyon, France; Paul Scherrer Institute, Switzerland (1984); St Petersburg, Russia; Moscow, Russia; Dubna, Russia.
  - Other Places: Seoul, Korea; Zibo, China (2004).

Hadron therapy was initiated at Berkeley in 1938 [17]. It developed rapidly, and there are now advanced centers of physics such as CERN in Geneva, which has the largest hadron accelerator in the world (the LHC). CERN also has the best school of hadron therapy in the world, with scholars attending from around the world, mainly ENTERVISION research fellows. Approximately one third of the 15,000 accelerators operating in the world are used in medicine: 3 % are employed in nuclear medicine and 30 % in radiation therapy. Most of them produce X-rays, while only 25 are used as beam sources of hadrons.

Tumors that are more sensitive to treatment using hadron therapy include chondrosarcomas, arteriovenous malformations, and uveal melanoma. It is important that a multidisciplinary team—consisting of medics, physicists, technicians, and others—is involved in research and treatment so that there is an efficient

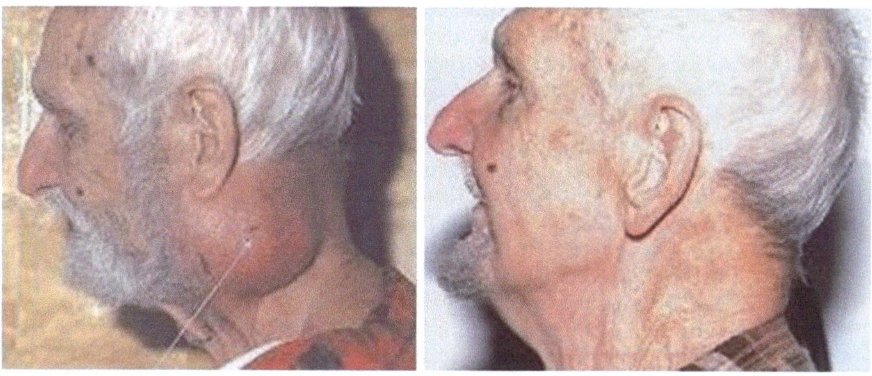

Fig. 1.23 Inoperable squamous cell tumor before and 2 years after fast neutron therapy [18]. Arlene Lennox (presented in a seminar at Fermilab on Nov 21, 2003) [15]

exchange of ideas. Hadron equipment is very expensive and this is perhaps the most important limitation in acquiring this resource throughout the world.

Two corroborative photographs demonstrating the treatment efficiency of fast neutron therapy for an inoperable tumor are shown in Fig. 1.23.

The neutron therapy unit at Wayne State University in Detroit, USA, called the Gershenson Radiation Oncology Center at Karmanos Center (KCC/WSU), has more experience than any other facility in the world regarding the use of fast neutron therapy for prostate cancer; they have treated more than 1,000 patients over the last 10 years. The KCC produces its neutron beam by accelerating deuterons with an energy of 48.5 MeV on to a beryllium target. The deuterons are accelerated using a super-conducting cyclotron, where the source of neutrons can spin 360° around the patient. Neutrons can be obtained by accelerating protons (p) or deuterium (2H) and making them collide with a beryllium (Be) or lithium (Li) target, thus provoking reactions of the following types: 9Be (p, n) 9B; 7Li (p, n) 7Be; and 3H (2H, n) 4He.

References

1. http://www.nobelprize.org/nobel_prizes/physics/laureates/1901/rontgen-bio.html
2. Mishima Y (ed) (1996) Cancer neutron capture therapy. Springer, New York
3. http://www.bestcyclotron.com/particletherapy/#
4. Charlie Ma C-M, Lomax T (eds) (2012) Proton and carbon Ion THERAPY (Imaging in medical diagnosis and therapy). CRC Press, Boca Rota
5. Amaldi U, Kraft G (2005) Radiotherapy with beams of carbon ions. Rep Prog Phys 68:1861–1882
6. Dosanjh M (2012) From physics to medicine: hadron therapy. CERN—Bulgarian Industry. http://nis-su.eu/Documents/Actions/Presentation_Manjit_Dosanjh.pdf
7. Jakob B, Splinter J, Taucher-Scholz G (2009) Positional stability of damaged chromatin domains along radiation tracks in mammalian cells. Radiat Res 171:405–418
8. https://accelconf.web.cern.ch/accelconf/r08/talks/THBAU02.pdf
9. http://www.cdc.gov
10. Ferlay J, Soerjomataram I, Ervik M, Dikshit R, Eser S, Mathers C, Rebelo M, Parkin D.M, Forman D, Bray F (2013) GLOBOCAN 2012 v1.0, Cancer Incidence and Mortality Worldwide: IARC CancerBase No. 11 [Internet]. Lyon, France: International Agency for Research on Cancer; 2013. http://globocan.iarc.fr. Accessed 20 Jan 2015
11. Tsujii H, Kamada T, Shirai T et al (2013) Carbon-ion radiotherapy: principles. Practices and Treatment Planning, Springer Japan
12. Wilson R (1946) Radiological use of fast protons. Radiology 47:487–491
13. Amaldi U, Bonomi R, Braccini S et al (2010) Accelerators for hadrontherapy: from Lawrence cyclotrons to linacs. Nucl Instrum Methods Phys Res A 620:563–577
14. Chao AW, Chou W (eds) (2009) Reviews of accelerator science and technology, vol 2. Medical Applications of Accelerators, World Scientific, Singapore
15. Nunes MA (2013) Hadron therapy physics and simulations. Springer, New York
16. Lawrence JH, Tobias CA, Born JL, McCombs RK, Roberts JE, Anger HO, Low-Beer BV, Huggins CB (1958) Pituitary irradiation with high-energy proton beams: a preliminary report. Cancer Res 18(2):121–134
17. www.slac.stanford.edu/slac/sass/talks/aiden_sass.pdf
18. http://www-bd.fnal.gov/ntf/reference/hadrontreat.pdf

Chapter 2
Equipment and Techniques

2.1 The Big Bang, Hadrons, and the Evolution of Energy

2.1.1 The Big Bang

How did energy and matter evolve immediately after the Big Bang? Approximately 13.7 billion years ago, an explosion called the Big Bang gave birth to the universe (Fig. 2.1), with the universe expanding and cooling. The Big Bang corresponds to the quantum epoch. The probable history of the universe was divided didactically into two eras: the era of radiation and the era of matter. The radiation era initially experienced the differentiation of quarks, forming the epoch of grand unification (10^{-34} s). The electroweak epoch (asymmetry) and the formation of nucleons (10^{-10} s) with quark confinement followed, forming protons and neutrons and experiencing the disappearance of anti-quarks.

At the end of this era of radiation, we entered the era of matter, relating to the formation of the nucleus (180 s) with decoupling matter-radiation, succeeded by forming atoms (300,000 years), and the universe became transparent. Finally, stars were formed (1 billion years) with the first supernova. Here, the formation of heavy atoms, protogalaxies, and black holes occurred. Approximately 13.7 billion years later is the present time, with the spiral galaxy and solar system, etc. [1]. Thus, after just 10^{-10} s from the Big Bang, protons were already formed.

For more information about the physics of the Big Bang related to hadron therapy, the reader is referred to the book *Particle Accelerators: From Big Bang to Physics Hadron Therapy* by Amaldi [2].

It is worth noting that only 5 % of the matter in the universe is visible matter (which is subject to the standard model). The rest is dark matter (20 %) and dark energy (75 %) [1], which does not apply this model (see Fig. 2.2). Our knowledge of the universe is minimal, but the advantage that the universe has over man—being immense and eternal—is not known to it because the universe does not think (as the French philosopher Pascal said).

M. d'Ávila Nunes, *Protontherapy Versus Carbon Ion Therapy*,
Biological and Medical Physics, Biomedical Engineering,
DOI 10.1007/978-3-319-18983-3_2

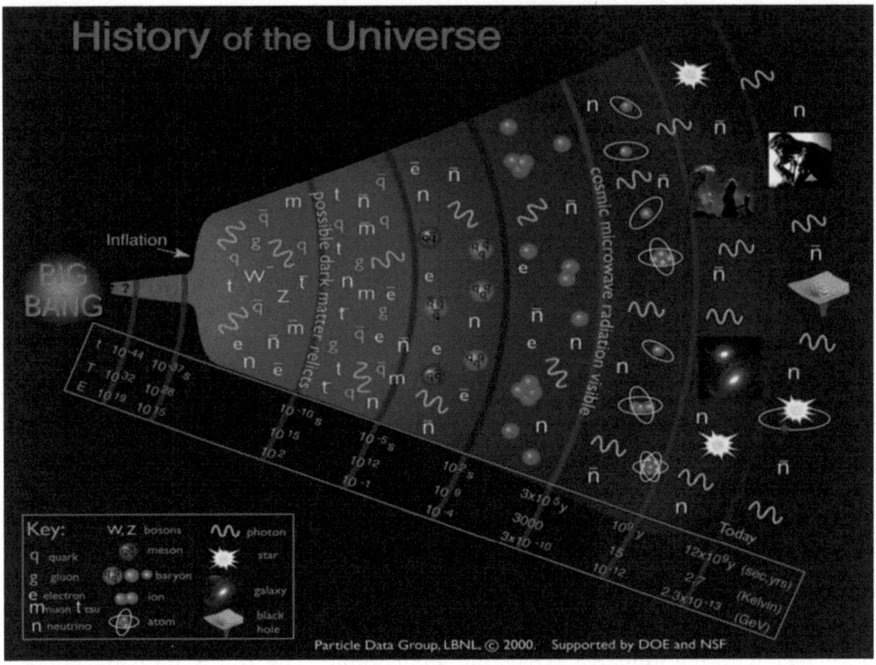

Fig. 2.1 Brief history of the universe. *Courtesy* Wikimedia Commons, [3]

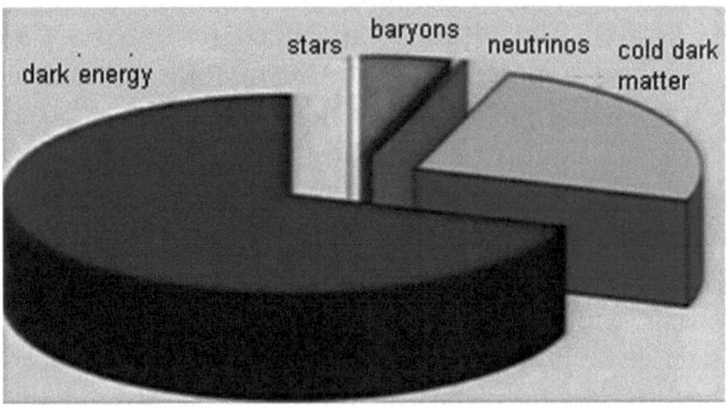

Fig. 2.2 Composition of the observable universe. Reproduced from [1]

2.1.2 Hadrons

The hadrons known to date are protons, neutrons, and a meson (pions). Pions are the most common types of particles in a particle collision and may be considered mild proton mass to approximately 15 % of that of the proton. There are two kinds of hadrons, classified according to their spins: baryons (half spin) and mesons (integer spin). Thus, the hadrons [1] are baryons (proton [p], neutron [n]) and mesons (pion π^+, π^- and π^o)

Hadrons penetrate deeper than electrons and photons but not as deeply as muons and neutrinos. In general hadron therapy, it is considered as hadrons protons, neutrons, pions, ions (alpha, C, N). The hadrons are highly interactive particles experiencing strong force. Hadron therapy (HT) is radiation therapy that uses hadrons. The strength of HT lies in the physical and radiobiological properties unique to these particles; they can penetrate tissues with limited diffusion and deposit maximum energy just before stopping (Bragg peak). This allows a precisely defined region to be specifically irradiated. HT allows access to a more controlled distance than conventional radiotherapy; however, the patient cannot move during application so that the radiation does not harm healthy tissue. Thus, with the use of hadrons, the tumor can be irradiated with less damage to healthy tissue compared with X-ray [4, 5].

2.1.3 *Evolution of Energy*

When working with particle accelerators, smaller distances are tested for higher energy accelerated particles. In 1930, testing values for a distance of 10^{-11} m required about 100 keV. Twenty years later, approximately 100 MeV was reached with testing distances of 10^{-14} m. In 1970, 100 GeV was reached and distances up to 10^{-17} m could be tested; in 1990, 10^{-18} m was reached with energy of 1 TeV. Currently, 10 TeV in the large hadron collider (LHC) has reached 10^{-19} m. How much we will we be able to reach in the coming years as accelerators grow in size, complexity, and cost?

Figure 2.3 shows a sharp increase in power over time: an order of magnitude for every 6–10 years. Each generation replaces the previous for increasingly high energy. It is important to note that energy is not the only interesting parameter: consider also the intensity and size of the beam.

Protontherapy uses accelerators with energy around 200–250 MeV/u, whereas carbon ion therapy uses 400–450 MeV/u with a current of 0.1 nA [1]. Circular accelerators, such as the Cyclotron and Synchrotron, are the most frequently used. The Cyclinac (which is a combination between a linac and Cyclotron accelerator) and the Laser and dielectric wall accelerators (DWA) are in development; if successful, they will reduce the cost, size, and complexity of current accelerators. The largest particle accelerator in the world, the LHC, has a biomedical facility for advanced research and education, called Low Energy Ionizing Ring (LEIR; http://medicalphysicsweb.org/cws/article/opinion/56295) to study basic physical and radiobiology, carbon ion fragmentation, dosimetry, and test instrumentation.

All circular accelerators use a linear accelerator), which, by means of electrical fields generated by radiofrequency (RF) cavities accelerate particles, and the bending trajectory, and the focusing, made by magnets. For the LHC to preserve proton beams in a path of 27 km, the magnetic fields require constant adjustment to compensate for the beam energy increases. The magnetic fields are on the order of

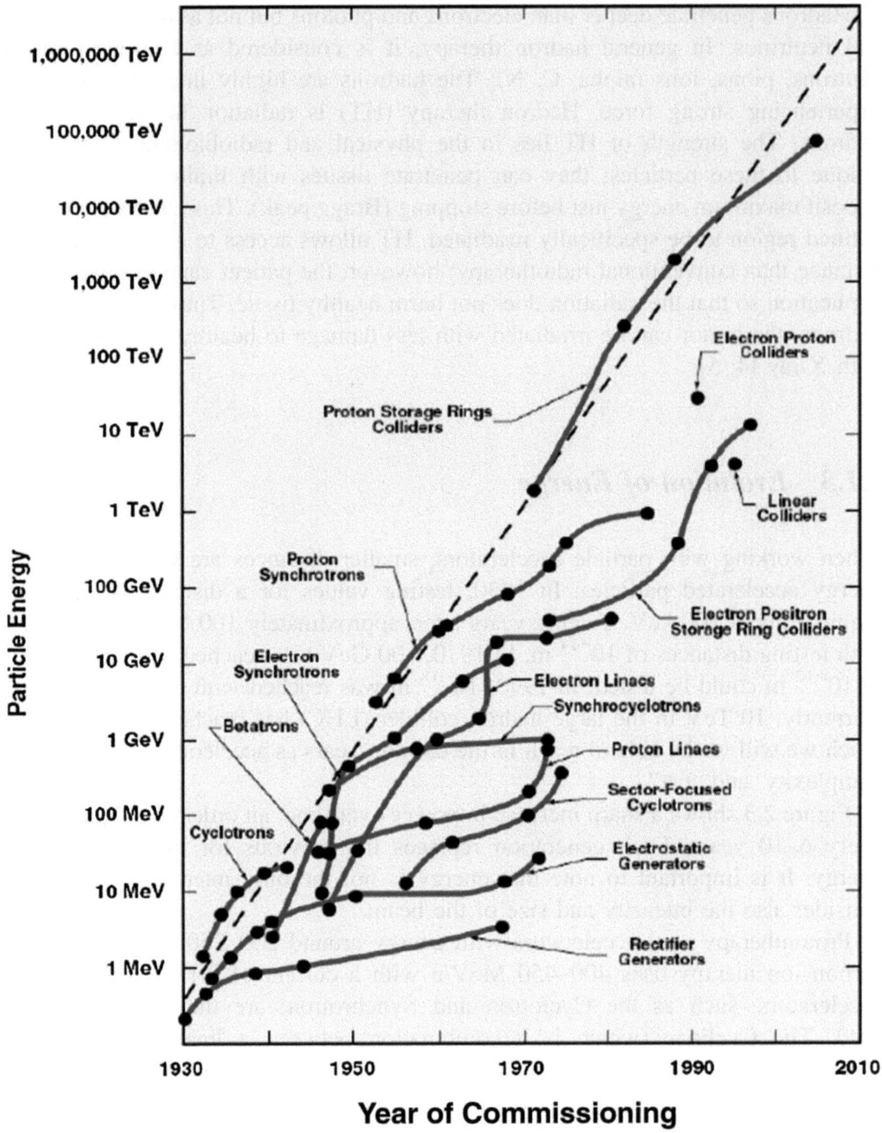

Fig. 2.3 A Livingston plot showing the evolution of accelerator laboratory energy from 1930 until 2005 [6]. The energy of colliders is plotted in terms of the laboratory energy of particles colliding with a proton at rest to reach the same center of mass energy. *Courtesy* Prof. Dr. Wolfgang Kurt Hermann "Pief" Panofsky

8.3 T—in other words, 170,000 times stronger than the earth's magnetic field. A field of this value requires a current of approximately 11,800 A. The 1232 magnets consume an electric current that could fill the needs of a small town with 150,000 houses!

Considering the LHC complex as a whole, the power consumption would be about 120 million watts of electrical power during peak demand [7]. The stored energy totals 11 billion joules—the amount of energy stored in 2.5 tons of TNT, spread over the 27 km. Each bending magnet of the trajectory of the protons is 14 m long and weighs 35 tons. The system temperature is maintained at −271 °C, requiring 10,800 tons of liquid nitrogen, followed by 120 tons of liquid helium (the cooling process takes a month and a half). The cables are made of an alloy of superconducting niobium-titanium. The length of the filaments used in bending magnets is equivalent to five times the distance from Earth to the sun and back, with plenty for any round trip around the moon (Don Lincoln). Thus, we can say that we have entered a golden age with new discoveries in all areas of human knowledge—physics, astronomy, chemistry, mathematics, computer simulation, industry, and others—through the creation of the LHC and the injection of $10 billion from participating nations in its construction.

Only with international integrated cooperation was it possible to build the LHC. With the additional experiments to be performed at the LHC, we believe that the importance may not be learning about elementary particles but to bring a deeper understanding of the fundamental questions about the universe. Finally, hadronic therapy will eventually move through physical and biophysical research; with the use of the LEIR of the European Organization for Nuclear Research (CERN), researchers will embrace new therapeutic approaches to save lives and cancer will be destroyed.

2.2 The Cyclotron, Ernest Orlando Lawrence, and Equations

The cyclotron was invented in 1932 by Ernest Lawrence. It accelerated protons with a fixed frequency of up to 1.25 MeV, allowing nuclear transmutation [8, 9]. Lawrence received the Nobel Prize 7 years later. The University of Berkeley recognized the potential of this new machine and built a 5-m-long cyclotron that accelerated protons to an energy of 20 MeV. Figure 2.4 shows two current cyclotrons and a schematic drawing in which the magnetic field imposes a circular path on the particles. The oscillating electric field (RF) is responsible for particle acceleration; the final trajectory is a spiral.

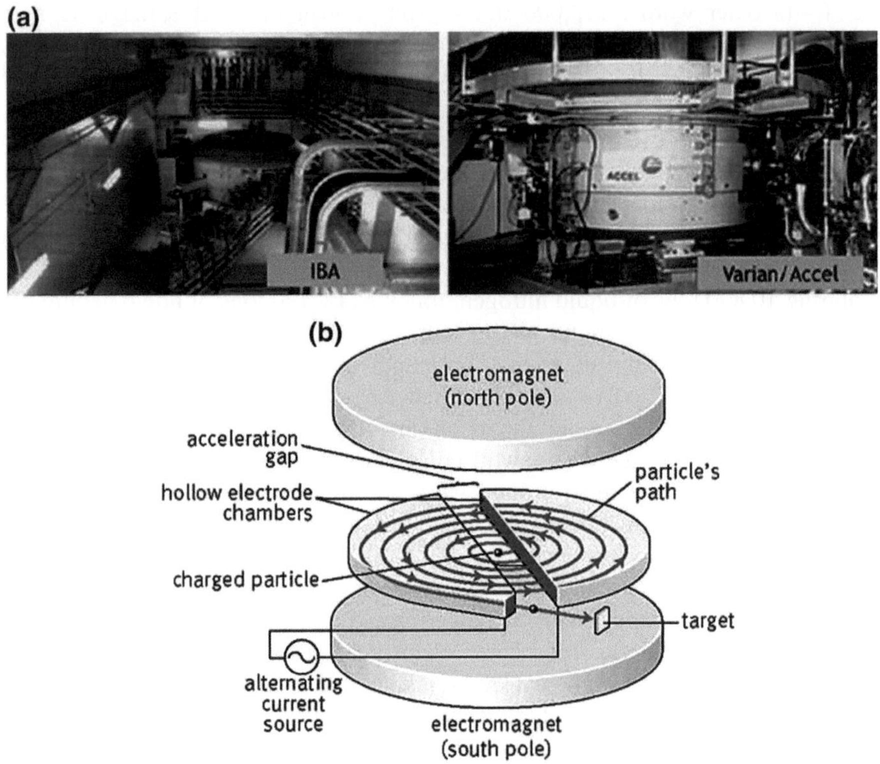

Fig. 2.4 **a** Commercial cyclotrons (Ion Beam Applications variants and the Varian Accelerator). **b** Schematic drawing of a cyclotron. Reproduced from [10]

2.2.1 Motion of Particles: Equations

The force acting on a particle, with velocity v in a magnetic field B, has the following characteristics. The direction is perpendicular to the plane (v, B); this can be represented by the lefthand rule where the thumb, index, and middle fingers are mutually perpendicular (90°). The thumb indicates the direction of the force, the index finger is the vector magnetic field, and the middle finger is the speed. The magnitude of the magnetic force that acts on the particle is given by $f = qv\, B\sin\, \phi$, where f is the magnetic force, q is the particle charge in Coulombs, and ϕ is the angle between the vectors B and v (this angle can vary from 0 to 180°). This is explained in more detail in Fig. 2.5.

Furthermore, when observing the particles from a proton–proton collision, for example, it can be concluded that particles of lower energy bend more and higher energy particles bend less because the magnetic field can bend the path of a particle. Because the magnetic force depends on the active particle charge, the trajectory can curve in either direction. If v is perpendicular to B, the equality decreases to $F = qvB$. However, Newton's equation gives the expression for the force F = mass

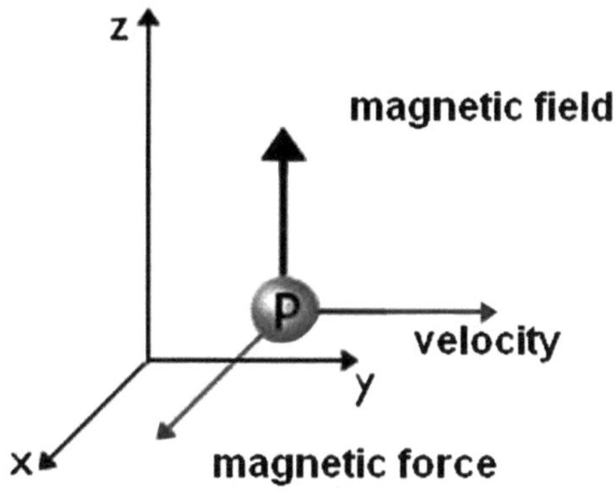

Fig. 2.5 Magnetic field vectors, velocity, and magnetic force acting on a proton [1]

(m) × acceleration (a), which means that to find the necessary replacements, the following equations are used for the speed and radius:

$$v = qBr/m \text{ and } r = mv/qB$$

Because we know v, we can calculate the acceleration as $a = v^2/r$. Substituting the value of v in the equation, the acceleration is obtained. Because the radiated power is given by $P\alpha\ (q^2\ a^2/c^2)$—that is, $P\alpha\ (qB/m)^4\ r^2$, it can be concluded that a smaller mass has greater radiated power, which causes the withdrawal of the beam particle. Thus, the electron radiates more power than the proton.

2.2.2 *Calculating the Frequency of the Cyclotron*

Calculation of the frequency of the cyclotron is very straightforward. The total turn is equal to $2\pi r$ in a path of radius r. If t is defined as the time spent in the half turn, $v = \pi r/t$, then $t = \pi r/v$. However, because $v = qBr/m$, then $t = \pi m/qB$. Thus, the time spent on the course is the same for all orbits, independent of the radius. Because the period of one complete turn (T) is twice that spent in the half turn $T = 2t$, then $T = 2\pi m/qB$. Because the frequency (ν) is the inverse of the period, we have $\nu = 1/T$, and thus $\nu = qB/2\pi m$. The angular frequency becomes $\omega = 2\pi\nu$, then $\omega = qB/m$.

This is the frequency value obtained from the RF source to produce the acceleration of a charged particle q and mass m, which are subjected to the magnetic field B. It can be concluded that the cyclotron frequency is directly proportional to B and inversely proportional to the ratio m/q. Thus, the particle with the lowest m/q ratio produces a spiral with more full turns (higher frequency), provided that the field remains constant.

A video on the cyclotron can be viewed at http://www.youtube.com/watch?v=cNnNM2ZqIsc.

If a cyclotron (200 MeV) were as small and inexpensive as the 5–20 MeV linacs used in conventional radiotherapy, then more than 90 % of patients could be treated with a proton beam. The accelerators used today are large and expensive, costing around 20 million Euros for a proton accelerator and 40 million Euros for a carbon ion beam facility. The installation of gantries would add another 10–12 million Euros to the cost. The gantry used at the Heidelberg Ion-Beam Therapy Center (HIT) weighs 670 tons and consumes 400 kW of power. Considering the LHC complex as a whole, the power consumption would be approximately 120 million watts of electrical power at peak demand. The stored energy is 11 billion joules (1). In the future, it is possible that gantries will be built using superconducting magnets. The current situation regarding size and costs is expected to change in the future. The Belgian company Ion Beam Applications (IBA) already offers a superconducting cyclotron with a 6 m diameter, which accelerates carbon ions up to 400 MeV/u. The TERA Foundation introduced and developed a new type of accelerator, the cyclinac [11], to accelerate protons and carbon ions, with a time of only 1 ms required to vary the energy of the beam, compared with the 20–50 ms needed by a cyclotron and the 1 s needed by a synchrotron.

2.3 The Proton Synchrotron, E.M. McMillan and V. Veskler

Conceptually, the principle of the synchrotron was published in a Russian newspaper by Vladimir Veksler; however, it was built by Edwin McMillan in 1945 (Fig. 2.6). The first proton synchrotron was designed by Sir Mark Oliphant, Australian physicist, and built in 1952 [12]. In particle physics, the synchrotron is an accelerator of cyclic particles in which the electric field is responsible for the acceleration of the particles and the magnetic field is responsible for the change of direction of the particles; both fields are synchronized with a beam of particles. The magnetic field is increased to keep the charged particles in a constant radius orbit as they reach higher speeds. Because the radius is constant, the "dees" used in the cyclotron are not needed and the particle moves in an annular chamber vacuum in a ring-shaped magnet. One or more resonant cavities are used to accelerate particles. An RF is applied into the cavity so that the particles are attracted when they approach and repelled when they leave it (Fig. 2.7).

The orbit of a synchrotron is not a circle; rather, straight sections are added by the RF cavities, injection, and extraction, etc. Usually, the pre-accelerated beam is a linac (or small synchrotron, prior to injection). The curvature of beam radius does not match the machine radius.

An interesting comparison was proposed by Don Lincoln in his book *The Quantum Frontier* [7], which facilitates an understanding of the functioning of a proton synchrotron. The principle governing this accelerator is the same as that governing a tetherball (i.e., a ball attached to one end of a rope, with the other end

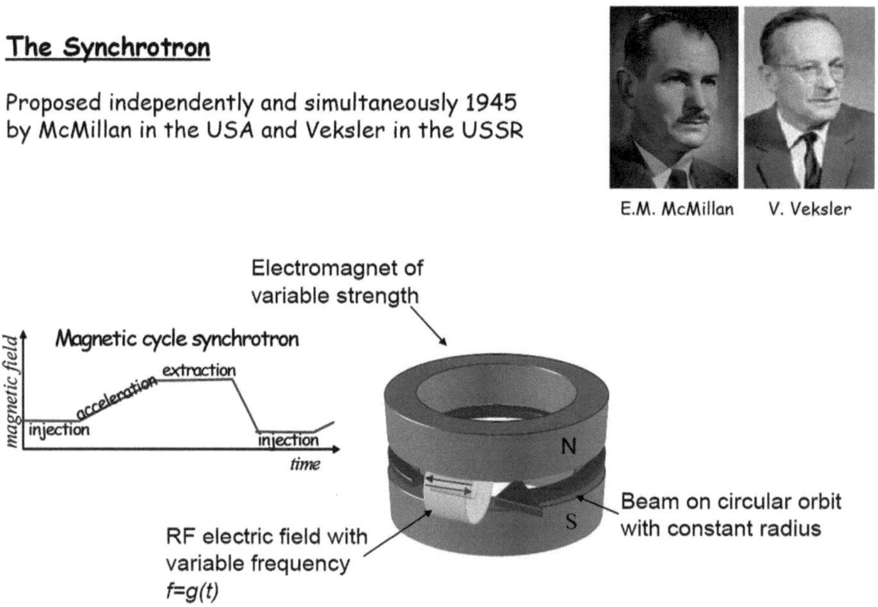

Fig. 2.6 The Synchrotron. *Courtesy* Prof. Dr. Hans-H. Braun

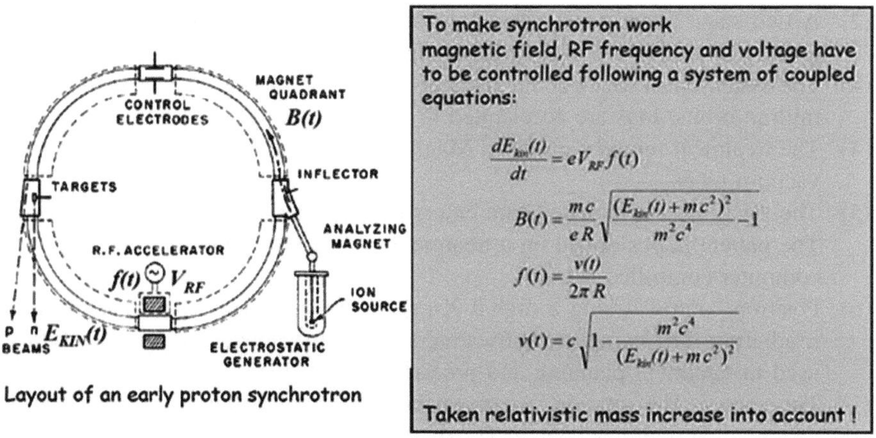

Fig. 2.7 Layout of an early proton synchrotron and a system of coupled equations. A synchrotron can be built for very high energies. For proton beams, limits are given by achievable magnetic field and size. *Courtesy* Prof. Dr. Hans-H. Braun

attached to the top of a tall pole that is anchored deep in the ground). A person hits the tetherball and the rope ensures that the ball travels in a circular path. Once the ball makes a full circle, it is hit again. The ball goes faster and makes another circuit. If the rope is attached to the top of the pole, it does not wrap itself around the pole; in principle, the ball can be made to travel very rapidly by synchronizing both its orbit and the person hitting it (hence the derivation of the name synchrotron). In a proton synchrotron, the electric field "hits" the proton and accelerates it. However, the counterpart of the rope in the tetherball analogy is not provided by electric fields, but rather by magnetic fields. Particles are accelerated by an electric field over a short distance and are then guided by magnetic fields in a circular path back to the electric field region for another round of acceleration.

The synchrotron was based on the cyclotron with a time-dependent magnetic guide, which was synchronized to a particle beam with increasing kinetic energy. The difference between the cyclotron and synchrotron is that the latter uses the principle of phase stability, maintaining the synchronism between the applied electric field and the frequency of revolution of the particle. Beam focusing and acceleration can be separated into different components using the curvature of the synchrotron beam. Thus, radiofrequency cavities are used for acceleration, the magnetic dipoles for particle deflection, and quadrupole/sextupole magnets for focusing the beam particles. The magnetic field maintains the orbit instead of accelerating the particles, and hence the magnetic field lines are only necessary in the region defined by the orbit.

The synchrotron facility consists of the following components (Fig. 2.8):

(1) The ion source accelerator: This is where ion beams composed of positively charged atoms are produced. For protons, hydrogen gas is used. For carbon ions, dioxide is used.
(2) A two-stage linear accelerator: Ions are accelerated in structures at high frequency up to 10 % of the speed of light.
(3) The synchrotron: Six 60° magnets bend the ions into a circular path. After a million orbits, ions are accelerated to 75 % of the speed of light.
(4) The treatment room beam lines: Magnets guide and focus the beam of ions in vacuum tubes.
(5) The treatment room: The beam enters the treatment room through a window. The patient is positioned on a treatment table that is adjusted accurately by a computer-controlled robot.
(6) Position control: Using a digital X-ray machine, images are obtained before irradiation. The computer software compares the images obtained with those used in treatment planning and precisely adjusts the position of the patient.
(7) The gantry: The rotation system enables the beam to be directed toward the patient at an optimized angle. The gantry weighs about 670 tons—600 tons of which can be rotated with submillimeter accuracy.
(8) The treatment room in the gantry: This is where the beam exits the gantry (beam line). Two rotation systems and digital X-rays are used to optimize the position of the patient guided by the images taken before irradiation.

The combination of magnetic field "guides", time dependency, and the principle of strong focus enables the design and operation of modern large-scale accelerators as colliders and even synchrotron light sources, such as at the Brazilian Synchrotron Light Laboratory in Campinas, São Paulo. The power limit could be increased by

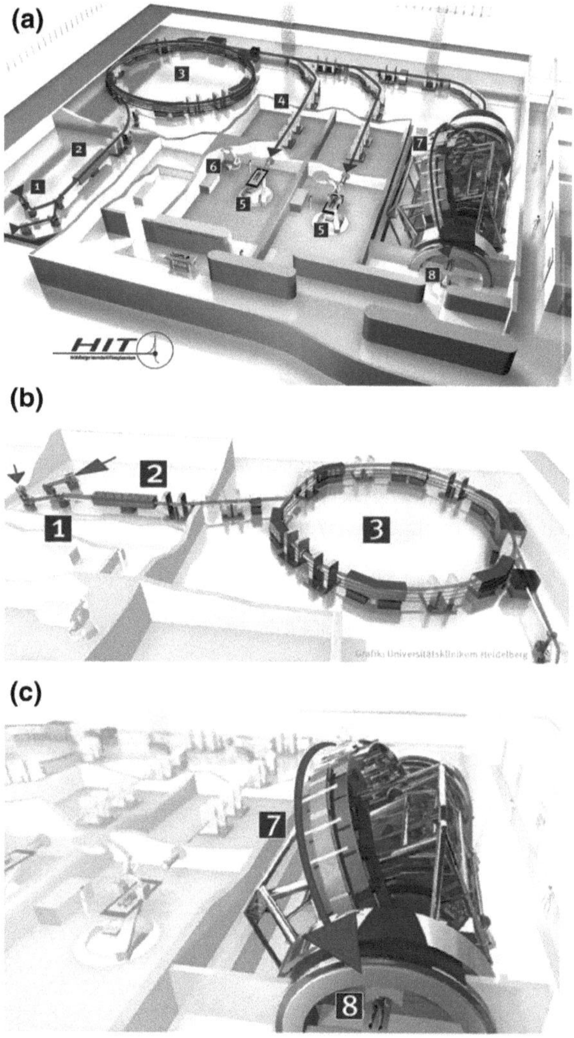

Fig. 2.8 **a** Synchrotron at the HIT in Heidelberg. **b** Schematic of the synchrotron at the HIT. **c** Schematic drawing of the HIT gantry. *Courtesy* Annette Tuffs, Head of Corporate Communications/Press Office, Heidelberg University Hospital

using superconducting magnets, which are not limited by magnetic saturation. Electron and positron accelerators may be limited by the emission of synchrotron radiation, resulting in a partial loss of kinetic energy of the particle beam. Therefore, the energy of electron and positron accelerators is limited by the loss of this radiation, which does not happen with proton or ion accelerators. The energy of these accelerators is limited by the strength of the magnets and the cost.

In the synchrotron, particle injection is pre-accelerated using a linac, microtron, or even another synchrotron, because synchrotrons are unable to accelerate particles from zero kinetic energy. The Tevatron at Fermilab was the largest collider in the world in 2008. It accelerated protons and antiprotons to 1 TeV and then collided them. The LHC has seven times this energy; accordingly, the proton–proton collisions occur at about 14 TeV. The LHC also accelerates heavy ions (such as lead) up to an energy of 1.15 PeV [13].

2.4 Hybrid Systems: C400 from IBA and New Synchrotron from Brookhaven National Laboratory

The C400 from Ion Beam Applications (IBA, Belgium), is being developed in partnership with the Joint Institute for Nuclear Research (JINR, Dubna, Russia). It is a cyclotron with superconducting coils that can produce 400 MeV energy, allowing the acceleration of protons and carbon ions. This cyclotron follows the current trend in the development of systems for providing protontherapy and carbon ion therapy in the same equipment—that is, hybrid systems. However, despite the IBA having developed a compact gantry for ProteusOne, it does not have a similar system for the C400. The C400, shown in Figs. 2.9 and 2.10, has a diameter of 6.3 m; its estimated main parameters are provided in Table 2.1.

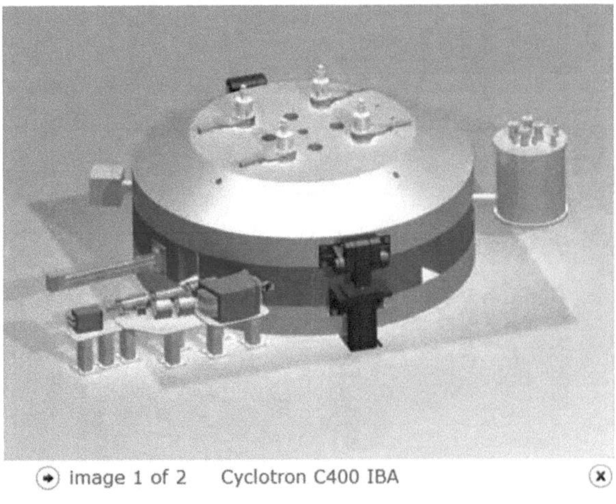

Fig. 2.9 Cyclotron C400 IBA. *Courtesy* IBA

Fig. 2.10 Cyclotron C400 IBA, about 6.3 m in diameter with superconducting coils. *Courtesy* IBA

Table 2.1 Estimated parameters of the C400 cyclotron in the pre-study phase

General properties	
Accelerated particles	H_2^-, $^4He^{2-}(\alpha)$, $(^6Li^{3+})$, $(^{10}B^{2-})$, $^{12}C^{6-}$
Injection energy	25 keV/Z
Final energy of ions, protons	400 MeV/u
	265 MeV/u
Extraction efficiency	~70 % (by deflector)
Number of turns	~2000
Magnetic system	
Total weight	700 t
Outer diameter	6.6 m
Height	3.4 m
Pole radius	1.87 m
Valley depth	0.6 m
Bending limit	K = 1600
Hill field	4.5 T
Valley field	2.45 T
RF system	
Number of cavities	2
Operating frequency	75 MHz, 4th harmonic
Radial dimension	1.87 m
Vertical dimension	1.16 m
dw voltage	
Center	80 kV
Extraction	160 kV

From [33]. *Courtesy* IBA

In January 2014 at the 5th Asian Forum for Accelerators and Detectors in Melbourne, Australia, the synchrotron Ion Rapid Cycling Synchrotron Medical (iRCMS) was presented (Fig. 2.11). The iRCMS was developed at Brookhaven National Laboratory through a cooperative research and development agreement with Best Medical International. The iRCMS will be used in future cancer therapy, working with protons and carbon ions. It was designed and optimized to provide maximum energy of 400 MeV/u and a frequency of 15 Hz for carbon ion therapy, effecting treatment at a maximum depth of 27 cm.

The iRCMS offers an advanced scanning spot with quick energy modulation, facilitating the release of beams with unprecedented accuracy. Its display is 12 m wide by 23 m length, and it uses one linac to inject protons and carbon ions to a kinetic energy of 8 MeV/u. Table 2.2 provides a comparison of several beam accelerators.

Because the iRCMS cycle is about 100 times faster than other "slow-cycling" synchrotrons, the number of protons accelerated per cycle can be as much as 100

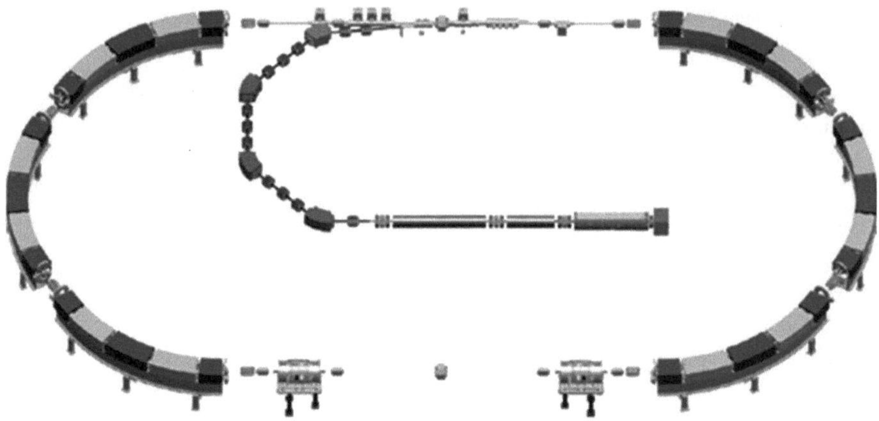

Fig. 2.11 Schematic diagram of the footprint of the medical synchrotron. The *rectangles* along the 180° arcs of 5 m radius are combined function magnets. The length of each of the straight sections is 12 m. The *bottom* straight section is dedicated to the RF acceleration system, and the *top* one is for the beam injection and extraction systems. The pre-accelerator that injects protons or C^{5+} ion bunches at 8 MeV/u is located in the area enclosed by the racetrack. Reproduced from: [14]

Table 2.2 Properties of the beam of various accelerators

Accelerator	The beam is always present?	The energy is electronically adjusted?	Which is the approx. time (in ms) to vary E_{max}?
Cyclotron	Yes	No	100
Synchrotron	No	Yes	1000
Linac	Yes	Yes	1

Reproduced from: [14]

times smaller for a fixed treatment time. This leads to five main advantages: faster energy change, less beams per cycle, efficient beam extraction, better control of delivered dose, and a smaller magnet size. Because less beams are used in the accelerator at any given time, it is far less likely that a worst-case incident would occur, in which excess beam is suddenly and inadvertently delivered to the patient. Low beam intensities also avoid the ravages of space charge effects, which at best cause the beam size to increase with intensity, and at worst put a hard limit on the intensity of the beam. Low beam intensities (per cycle) also allow the beam to be extracted from the synchrotron in a single turn of the accelerator, at an energy that can be easily modified from one cycle to the next [14].

Figure 2.12 provides a comparison of intrinsic spot width to protons (~206 MeV/u) and carbon ions (~400 MeV/u). Under these conditions, the minimum voxel volume would be 715 mm^3 for protons and 13.8 mm^3 for carbon ions, at a depth range up to 27 cm. Therefore, carbon ions are 52 times more accurate than protons.

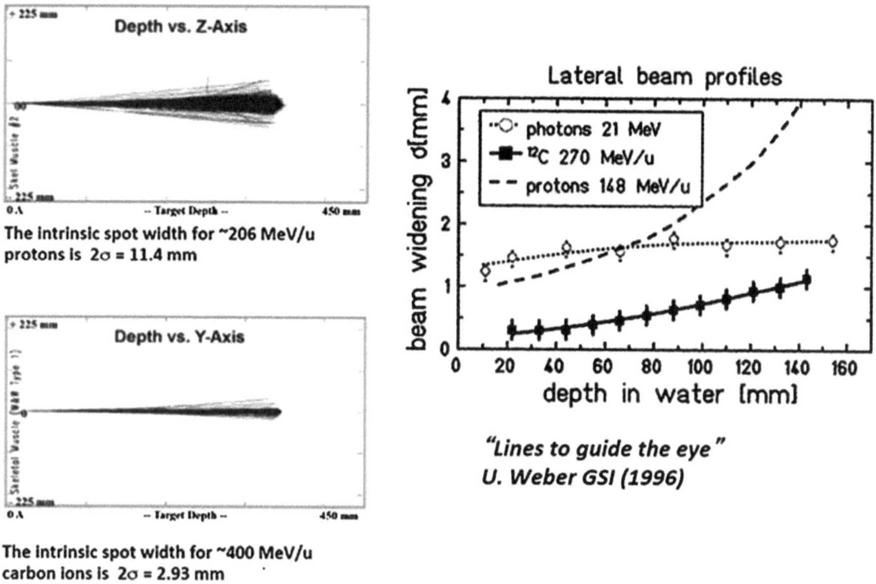

Fig. 2.12 Comparison of intrinsic spot width for protons and carbon ions. Reproduced from [14]

The beams of particles extracted from the synchrotron iRCMS are initially so small that the size of the final spot in the end of its range is just the intrinsic width given by the inevitable multiple scattering. The width of the intrinsic spot to protons at 206 MeV/u (at 270 mm depth) is 11.4 mm. The minimum value of the voxel volume is 715 mm^3. The intrinsic width of the spot for carbon ions at 400 MeV/u (at 270 mm depth) is 2.93 mm. The minimum value of the voxel volume is only 13.8 mm^3. Thus, carbon ions are 52 times more accurate than protons.

2.4.1 Summary

Best Medical International and Brookhaven National Laboratory are jointly developing a rapid-cycle proton/carbon synchrotron that enables advanced features, including a unique combination of advanced spot scanning with rapid energy modulation and elimination of the neutron contamination associated with patient-specific hardware. This technology has many advantages, including the following:

1. Intrinsically small beam emittances facilitating beam delivery with unprecedented precision
2. Small beam sizes, small magnets, and light gantries
3. Highly efficient, single-turn extraction
4. Efficient extraction, less charge per bunch, and less shielding
5. Flexibility in the choice of protons or carbon, future beam delivery modalities.

2.5 Gantry Specifications: Compact Gantry

The gantry is an extremely useful, essential piece of equipment (Figs. 2.13, 2.14, 2.15 and 2.16). If a conventional room was used instead of a gantry, the radiation would have to be provided in horizontal, vertical, or inclined beams, which prevents proper treatment, including for tumors in children.

Companies selling equipment for protontherapy and carbon ion therapy offer gantries that allow irradiation at any angle. Only horizontal irradiation is used in the treatment of cancer in the eye or head and neck. For the torso, it is possible to note a change both in dimensions as the position changes (e.g., due to breathing), thus necessitating the technique of "gating." It also uses the vertical irradiation. Robotic systems to keep the patient in different positions are being developed.

Fig. 2.13 The Heidelberg carbon-ion gantry (18 m × 7 m, 600 ton). *Courtesy* Thomas Haberer/Heidelberg Ion Beam Therapy Centre. *Courtesy* Prof. Dr. Thomas Haberer, Head of HIT and MIT

Fig. 2.14 IBA proton gantry (diameter > 6 m, weight ~ 100 tons, proton beam displacement from isocenter <1 mm). *Courtesy* IBA

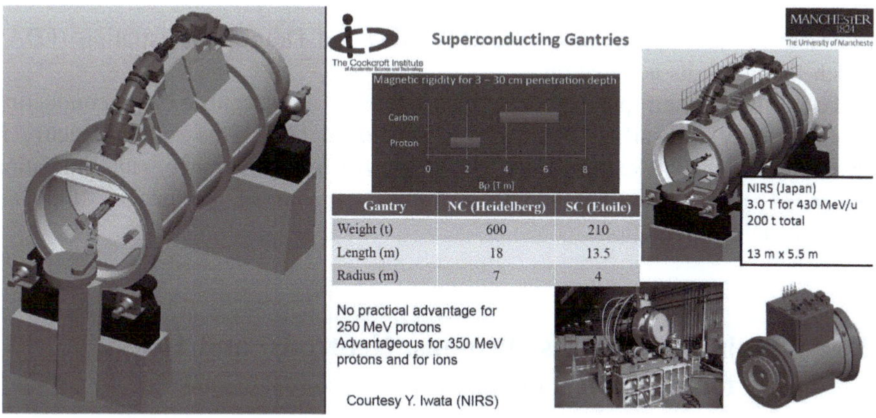

Fig. 2.15 Three-dimensional image of the NIRS superconducting rotating gantry for heavy-ion therapy (dimension 13 m × 5.5 m, weight 200 tons, 3.0 T for 430 MeV/u). *Courtesy* Prof. Dr. Y. Iwata, National Institute of Radiological Sciences (NIRS)

A hadron therapy center typically has 4–5 rooms. Researchers are working hard to create more affordable systems for the acceleration of protons and carbon ions. The IBA has adopted a more compact gantry system with a lower price, coupled to

Fig. 2.16 The gantry mechanism and final dipole of Gantry 2 at the Paul Scherrer Institute. The upstream design enables parallel scanning, which thereby simplifies treatment planning and gives an infinite SAD, which assists in skin sparing but necessitates a rather large 45-ton final dipole to give sufficient aperture to deliver the desired treatment field size. *Courtesy* Prof. Tony Lomax/PSI

ProteusONE, and approved by the U.S. Food and Drug Administration (FDA), which facilitates its acquisition.

To further obtain precise dose distributions, an isocentric superconducting rotating-gantry for carbon therapy was developed [15]. This rotating gantry is designed to transport carbon ions of 430 MeV/u to an isocenter with irradiation angles of more than ±180°. It is further capable of performing the fast raster-scanning irradiation.

2.6 Obtaining Particles (Protons, Neutrons) and Heavy Ions for Hadron Therapy

This section explains how to prepare protons, neutrons, and heavy ions for use in hadron therapy.

2.6.1 How Are Protons Obtained?

Protons are produced by applying an arc discharge of hydrogen gas into a source called a duoplasmatron. The electron is released from the hydrogen atom, leaving the positive nucleus, a proton, floating freely in the resulting plasma. By applying a strong electric field, the protons are extracted from the plasma surface and are sent on their way as a stream of positive particles. Currents of up to 300 mA can be obtained.

The protons interact with matter in three distinct ways:

1. They slow down through collisions with atomic electrons and finally stop.
2. They are deflected by collisions with atomic nuclei causing scattering.
3. Collisions with a nucleus yield secondary particles in motion. This is called nuclear interaction.

The first two conditions occur via electromagnetic interaction between the charge of the proton and the charge of the electrons or the atomic nucleus. There are mathematical theories for the first two conditions. Nuclear interactions are known to be infrequent and function according to a set of models. Even though computer programs are used to solve these problems, which are accurate to within seconds, predicting the dose for a patient is very complex and time-consuming.

2.6.2 How Are Neutrons Obtained?

Neutrons can be obtained by accelerating deuterons with an energy of 48.5 MeV onto a beryllium target. The deuterons are accelerated using a superconducting cyclotron. Generally, neutrons can be obtained by accelerating protons or deuterium and colliding them with a beryllium or lithium target, provoking reactions of the type 9Be (p, n) 9B, 7Li (p, n) 7Be, 3H (2H, n), and 4He (where p = proton, 2H = deuterium, n = neutron, Be = beryllium, B = boron, and 4He = helium-4).

James Chadwick discovered the neutron in 1932 using alpha particles (from radioactive polonium), with which he bombarded a blade of beryllium. He noted that uncharged particles left the beryllium bulwark. He placed in its path a paraffin bulwark from which the protons were discharged after bombardment by particles without being charged. The neutron was discovered! Its mass was determined in a way that was very similar to the proton because the impact removed protons from paraffin. The discovery of the neutron triggered a considerable increase in knowledge regarding nuclear structure. Additionally, in 1932, Werner Heisenberg realized that the nuclei of atoms were composed of protons and neutrons. He described the quantum mechanics involved and received the Nobel Prize in Physics in 1932. James Chadwick also received the Nobel Prize in 1935 for his discovery of the neutron. After the discovery of the neutron, Robert Stone began clinical trials with fast neutrons (radiation therapy) at the Lawrence Laboratory in Berkeley, CA, USA.

2.6.3 How Are Heavy Ions Obtained?

Heavy ions are atomic nuclei that have lost their electrons and are heavier than protons (hydrogen nuclei). A variety of ions are used, such as helium, carbon, and oxygen nuclei. Heavy ions are three times more effective than protons and helium ions. In the human body, heavy ions can be targeted with millimeter precision and are therefore superior to protons in the treatment of certain tumors. As is well known, ions are charged atoms. Thus, to obtain ions, atoms must necessarily lose their negatively charged electrons. For this purpose, carbon dioxide gas flowing within an ionic chamber is used. Free electrons in the gas are accelerated using magnetic fields and microwaves. Traveling through the ionic chamber, the electrons impact the molecules of carbon dioxide. After a collision, the molecules dissociate, and four of the six electrons in the carbon atom are separated. Electric fields are then employed to extract the carbon ions from the chamber. A special magnet transports them in a vacuum in a steady flow. This flow is converted into a pulsating flow with a frequency of 217 million pulses per second. The beam is collimated and the ions are accelerated. Subsequently, electromagnetic fields accelerate the ions to more than 10 % of the speed of light. Leaving the accelerator through a sheet of carbon, the carbon atoms lose their last two electrons, so that only nuclei with six positive charges remain.

2.7 Other Techniques in Development: Cyclinac, Laser, Dielectric Wall Accelerator

2.7.1 Cyclinac

The name *cyclinac* is a combination of cyclotron and linac. The cyclinac consists of a linac with a high frequency and a fast cycle that increases the energy of the particles previously accelerated by a cyclotron. The cyclinac can easily accelerate currents of the order of 2 nA, which are required for proton beam therapy (carbon ions), producing optimized ion beams for the irradiation of solid tumors using the most modern techniques [11]. The accelerators used for protontherapy are cyclotrons that are 4–5 m in diameter and synchrotrons of 6–8 m diameter. For carbon ion therapy, only synchrotrons of 20–25 m diameter are employed. Recently, large superconducting cyclotrons have been built for carbon ion acceleration.

Cyclinacs are excellent accelerators for hadron therapy because of the following characteristics: they work with frequencies of 300 Hz, allowing efficient tumor mapping; they have a low power consumption of 800 W (reducing costs); they rapidly modulate the active energy (1 ms) that is essential for studying organs in motion; and finally, they have steep acceleration gradients and have a reduced size (a cyclotron weighs 190 tons and a linac is 24 m long). The compact cyclotron used for the first particle acceleration to 120 MeV/u is smaller than the more widely used

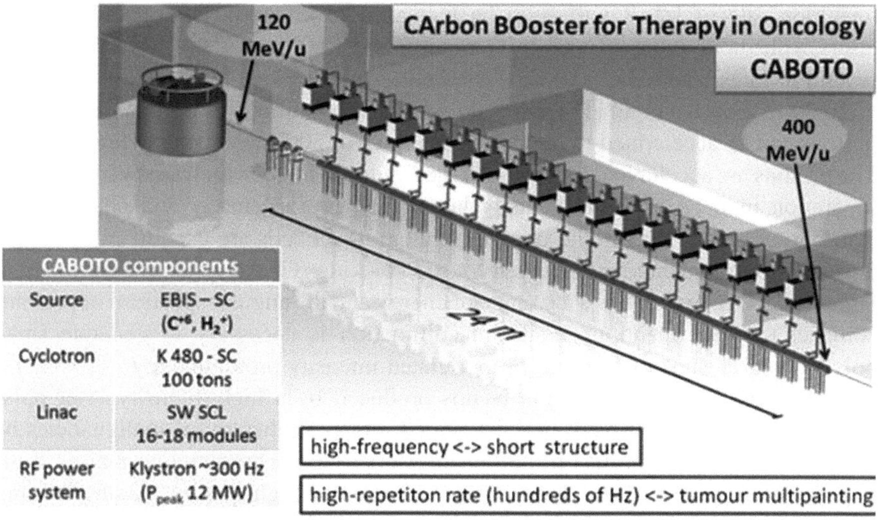

Fig. 2.17 The TERA Foundation's proposed cyclinac [11]. *Courtesy* Prof. Dr. Ugo Amaldi, President of Tera Foundation

cyclotron for protontherapy (e.g., IBA C235). Figure 2.17 shows the cyclinac (CABOTO) machine proposed by the TERA Foundation.

2.7.2 Use of Lasers in Hadron Therapy

Improvements in heavy ion therapy can be achieved using less expensive acceleration technologies. For potential use in therapy, the laser pulses need to accelerate protons to energies of ≤150 MeV and carbon ions to an energy of 350 MeV. Working with proton beams and focusing them is costly and difficult. Eccentric and isocentric equipment is used to transport proton beams from the final section of the accelerator to the target tumor. These structures are made of heavy magnets used to deflect the beam; they weigh 100–200 tons and have a diameter of 4–10 m. All of this equipment can generate costs of up to 150 million Euros.

Petawatt class lasers are not necessarily much smaller than the conventional accelerators used for hadron acceleration. The targets used to generate protons are only a few centimeters in size. Therefore, the target is positioned close to the patient, using small mirrors to transport the laser instead of heavy and expensive magnets. This makes the equipment much lighter, smaller, and less expensive. The laser beam can be sent to different treatment rooms using mirrors. In addition, the safety system for the laser is simple and inexpensive, and damage to the eyes of the doctor and the patient is prevented. An additional advantage is that the laser accelerator does not require radioprotection with thick concrete walls. The repetition rate of lasers will soon be increased to kilohertz levels, and petawatt lasers with diode-pumps (used in Germany) and fiber lasers will have power and repetition rates >100 Hz.

In 2002, Fourkal et al. [17] showed that, under conditions of optimal interaction, protons can be accelerated to relativistic energies of 300 MeV using petawatt lasers. The protons are accelerated by means of the Coulomb force, which arises from charge separation induced by high-intensity lasers. The proton energy and phase spatial distribution obtained from the particle simulations in cells are used to calculate the dose distribution using the GEometry ANd Tracking (GEANT) Monte Carlo simulation code. Because of the wide range of energy and the angular spectrum of protons, compact particle selection and beam collimation is necessary to generate small polyenergetic beams of protons for modulated-intensity protontherapy.

Intense and collimated proton beams produced by a high-intensity laser pulse interacting with plasma were first developed for protontherapy of malignancies by Bulanov et al. [16]. The fast proton beam was produced by directing a laser at the target, which generated accelerated proton beams of high quality. A simple comparison between the traditional accelerators and laser accelerators shows the superior qualities of the laser.

2.7.3 *How Are Protons Accelerated with a Laser?*

It is possible to accelerate protons by means of a violent acceleration of electrons in the laser field that draws protons behind them on the posterior surface of the target (Fig. 2.18). This creates a continuous proton spectrum. Computations have shown that by using two appropriately shaped targets, a scattering energy of 3 % can be achieved.

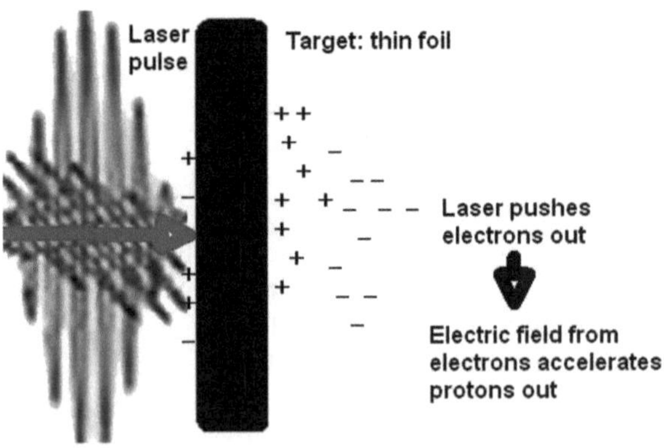

Fig. 2.18 Schematic drawing of a powerful laser blast onto a thin solid blade doped with hydrogen. Reproduced from [10]

In Fig. 2.18, a powerful laser pulse is shown acting violently on a target constituted by a thin blade doped with hydrogen atoms [18]. The laser accelerates electrons off the posterior region of the target, creating an electric field that favors the output of protons from the target. In the future, it is hoped that laser pulses with intensities in the range of 1018–1020 W/cm^2 and pulse durations of 30–50 fs will be possible [19]; this will allow a facility for treatment with protons (single facility) to be constructed based on illumination of a thin target.

Some companies are working to reduce the size and cost of high-power lasers, and there are several projects focused on improving beam quality. It is possible that the cyclotron will eventually become obsolete and will be replaced by more compact laser systems. Many years of dedicated research are needed to achieve this goal; for now, it is not considered to be economically advantageous.

Dielectric wall accelerators (DWA) are a type of induction accelerator. A traveling high-gradient field is created by switching high voltages on electrodes that are sandwiched between high-gradient insulators. This is the operating principle of the DWA. The accelerator tube is made of fused silica (250 μm thick), which is a pure transparent quartz and acts as an insulator. This method maximizes the electric field through the use of new insulators in the accelerator structures. However, with normal insulators, the maximum electric field strength is limited by the formation of sparks, which arise because the electrons repeatedly bombard the surface, creating an avalanche of electrons. Thus, to obtain a strong field accelerator, the formation of sparks must be presented by shortening the time during which the field is present. To decrease the insulation, the conventional high-voltage pulse of 1,000–1 ns leads to an increase in the surface cracking field of 5–20 MV/m. A new dielectric insulating configuration has enabled this limit to be increased to 100 MV/m. This high-gradient insulator (HGI) is constructed using a row of floating conductors sandwiched between sheets of insulators. Thus, a DWA can be made by forming rings of HGI-material and additional conductive sheets at frequent intervals along the stack. Each of these blades is connected to a high-voltage circuit with a switch. When these switches are closed, an electric field is produced in the inner side of the HGI ring. By successive closures of the switches along the stack, the region of the strong electric field is changed along the stack, and protons traveling in phase with the wave will be accelerated through these rings. This arrangement accelerates protons to 200 MeV in a system that is 2 m in diameter (Fig. 2.19).

Use of the DWA would circumvent some of the problems associated with conventional accelerators, such as their expense and enormous size; the DWA costs US $20 million and is much smaller than the accelerators used in the medical field. However, the DWA currently requires several improvements because of the high energies involved, including improvement of the high-gradient insulators. As compared with other proton accelerators, the DWA is apparently the only accelerator for which the power, intensity, and beam spot size can be varied pulse by pulse. The cyclotron only allows variation of the intensity under these conditions and the synchrotron allows variation of the energy and intensity, but not the spot size. The DWA allows variation of all these factors, pulse by pulse. Another advantage of this small linac is that it would be possible to mount it on a

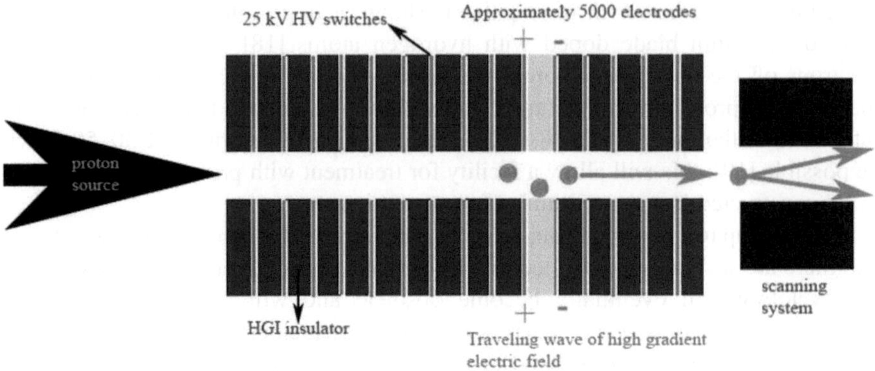

Fig. 2.19 Operating principle of the dielectric wall accelerator. Reproduced from: [10]

tomotherapy system. The Compact Particle Acceleration Corporation (CPAC) is developing a very flexible compact system for protontherapy based on the DWA that is much smaller and more powerful than conventional accelerators. The idea of a compact proton accelerator comes from a team led by George Caporaso of the Livermore Beam Research Program at the Physical and Life Sciences Directorate. Their HGI is built with layers made of metal, such as stainless steel, alternating with layers of insulating plastic, such as polystyrene.

An induction accelerator formed by a set of HGIs can maintain extreme voltages. A particle injector starts the action, and the transmission lines made of dielectric materials and embedded conductors produce the electric field that drives the particles along the tube. The transmission lines are called Blumleins (named after British inventor Alan Blumlein). A laser supplies power to switches in the Blumleins through a distribution system consisting of optical fibers. The small, solid-state silicon carbide optical switches open and close at high speeds to control the high voltage that reaches each Blumlein, increasing the energy of the particles as they traverse the tube. The opening and closing of each switch creates a virtual traveling wave that pushes the energized particles along the tube (Fig. 2.20).

This advance in size and power is due to three inventions: (1) the high-gradient insulator, which allows a substantial increase in voltage-holding capacity; (2) the optical switches, which can handle high-power loads at high speeds in a very compact size; and (3) the dielectric materials with embedded nanoparticles, which facilitate the transmission and isolation of extremely high voltages (Fig. 2.21).

Thus, the CPAC estimates that it will be able to create a system for protontherapy that is accessible to all cancer treatment centers and their patients. The DWA can be used to accelerate electrons, protons, or any ion, but more time is required before a clinical system can be established. Figure 2.22 shows a representation of the DWA.

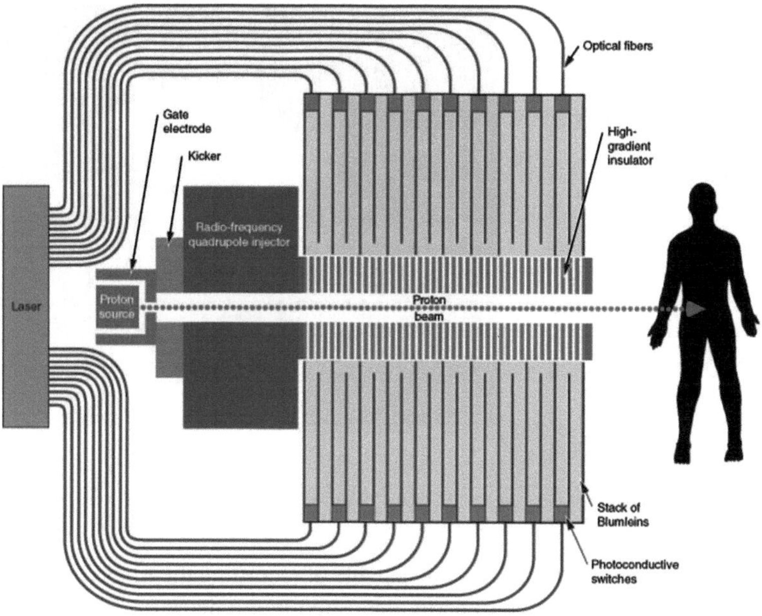

Fig. 2.20 Diagrammatic representation of the flow of protons in the dielectric wall accelerator from their source to the patient. Protons are sent to the interior of a "kicker" that injects them in pulses into a radiofrequency quadrupole, which compresses them into small bunches. Switches along the accelerator open and close at high speeds to control the voltage and increase the energy of the particles. Careful control of the switch mechanism creates a beam pulse with the velocity, shape, amplitude and length required for a given patient. Reproduced from [10]

Fig. 2.21 George J. Caporaso examines the Compact Particle Acceleration Corporation's (CPAC's) newest Blumlein design. Tests at the CPAC are combined with computer simulations at Livermore Laboratory to produce a practical design. *Courtesy* George J. Caporaso

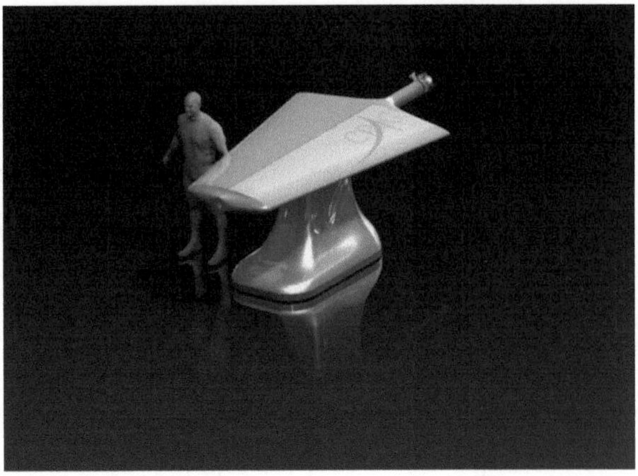

Fig. 2.22 An artist's rendition of the DWA in its fully developed form. *Courtesy* Dr. Anthony Zografos, Chief Operating Officer, Compact Particle Acceleration Corporation

2.8 Phantoms

The Timepix detector [20] is a system capable of recording the traits of characteristic particle shapes, including their energy deposited in the detector. The data recorded for each event allow estimation of the particle type, energy, and direction of flight. Opalka et al. conducted experiments for the detection and characterization of secondary radiation beams generated by primary therapeutic tissue equivalent material (water). The measurements were made in a water phantom irradiated by carbon ions in the Heidelberg Ion-Beam Therapy Center. Figures 2.23 and 2.24 show the pixel detector Timepix and the assembly for measures in a phantom water tank.

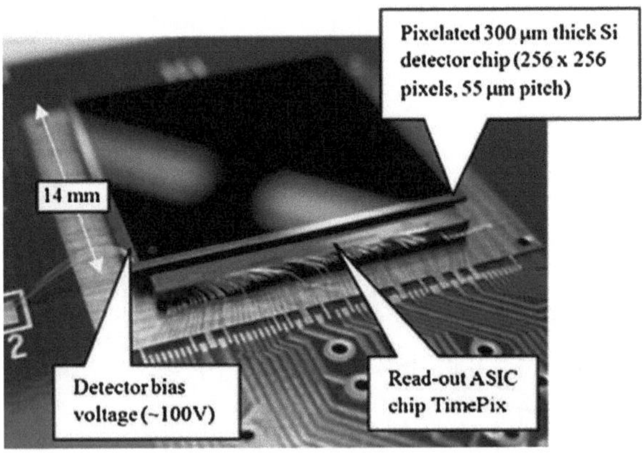

Fig. 2.23 The pixel detector Timepix. The device consists of two chips connected by a bump-bonding technique. The *upper chip* is a semiconductor sensor (usually silicon). The *bottom chip* is an ASIC readout containing a 256 × 256 matrix of preamplifier comparators and counters. *Courtesy* Prof. Dr. Lukas Opalka

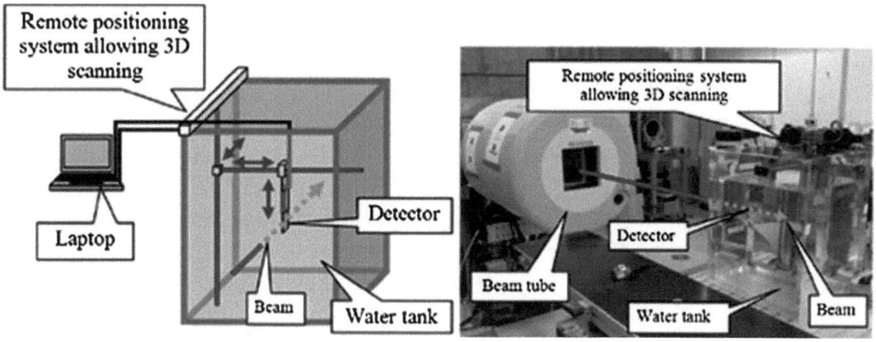

Fig. 2.24 Setup for measurements in the water tank phantom. The Timepix device and the readout interface were immersed in water inside using a waterproof rubber sleeve and mounted onto a three-dimensional positioning system. The beam axis is marked by the *red arrow*. *Courtesy* Prof. Dr. Lukas Opalka

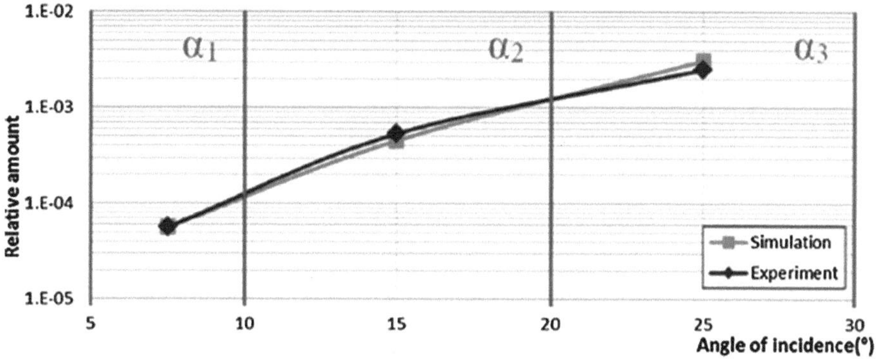

Fig. 2.25 A comparison of the measured number of secondary protons identified by Timepix and the prediction by Monte Carlo simulation at three angular intervals. *Courtesy* Prof. Dr. Lukas Opalka

It is also possible to achieve with this mount using a Monte Carlo simulation. Figure 2.25 shows a comparison between the number of secondary protons and the prediction by Monte Carlo simulation, which are in excellent agreement.

With such a system, it was possible to show that there are many events occurring beyond the Bragg peak, corresponding to secondary energetic particles, produced in the fragmentation processes as protons, fast neutrons, and gamma rays.

2.8.1 Microdosimetry Measurements

A tissue-equivalent proportional counter (TEPC) is a plastic sphere with a wall thickness of 1.27 mm and an internal diameter of 12.7 mm, which is equivalent to a few microns of a tissue sphere (Figs. 2.26 and 2.27). This device simulates a cell nucleus. The lineal energy is given by $y = \varepsilon/\lambda$, where y is the lineal energy, ε is the energy deposited in the TEPC, and $\lambda = 2/3d$ is the average length. In microdosimetry experiments, the lineal energy is measured using a TECP (Fig. 2.28).

The spheres are created to simulate cell nuclei. They have 1.27-mm-thick plastic walls, an inner radius of 12.7 mm, and are placed in water (phantom). The TEPC can be displaced on the translation table for any point XYZ in three dimensions (spatial). The measurements are performed by the electronics connected to the TEPC. To calculate the absorbed dose of radiation, the following equation can be used:

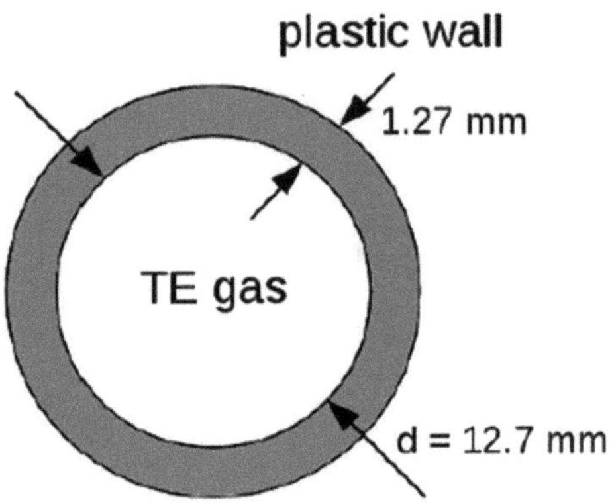

Fig. 2.26 Plastic sphere filled with gas at low pressure. TE = tissue equivalent. Reproduced from [10]

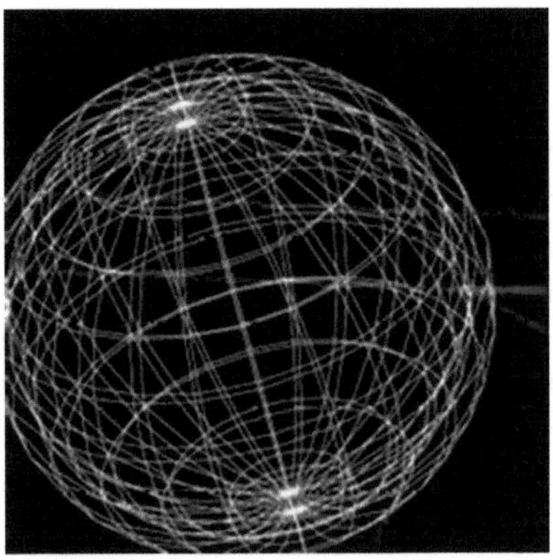

Fig. 2.27 *Sphere* with traces representing nucleons and nuclear fragments (*blue*). Fast electrons are shown in *red*. Reproduced from [10]

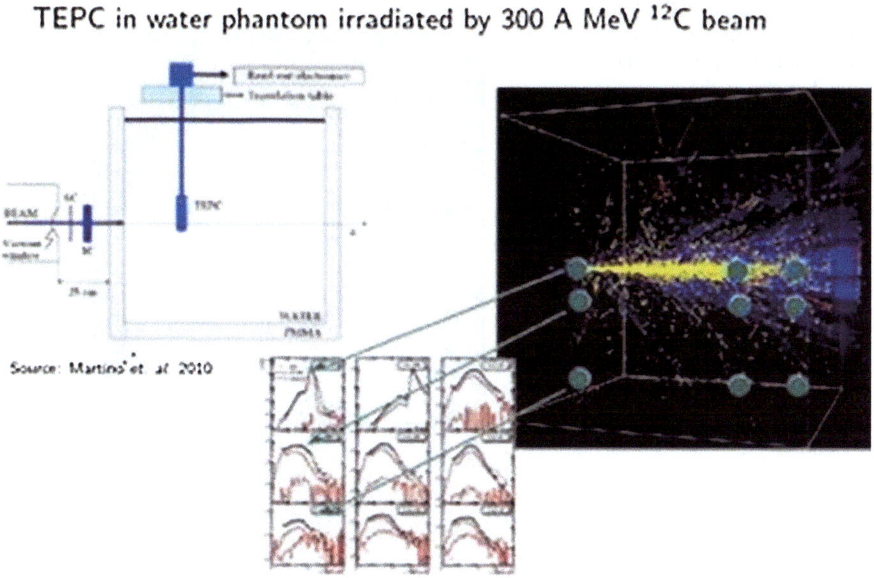

Fig. 2.28 Experimental arrangement used for microdosimetry at Gesellschaft fur Schwerionenforschung. Reproduced from [10]

$$D(\mathrm{Gy}) = \left(0.204/d^2\right)Y_f,$$

where d corresponds to the diameter of the simulated volume (μm), Y_f is given in keV/μm (average value), and D is the dose given in Gray (Gy). This unit is used in physical and not biological measurements.

2.9 Fluka: A Simulation Code

The simulation code FLUKA was originally designed for physics research involving accelerators and detectors. (See [21–26] for links that provide open access to FLUKA.)

Physicists use FLUKA to precisely predict electromagnetic and nuclear interactions in matter. For example, at CERN it is used to study beam-machine interactions and radiation damage. NASA has used it to analyze the radiation exposure of astronauts [27]. FLUKA is now used in state-of-the art therapy involving ion beam facilities, such as the HIT in Germany, to support treatment planning for cancer patients undergoing radiation therapy.

FLUKA used employed to generate large amounts of data and provide access to commercial software for treatment planning. It is also used for recalculating and

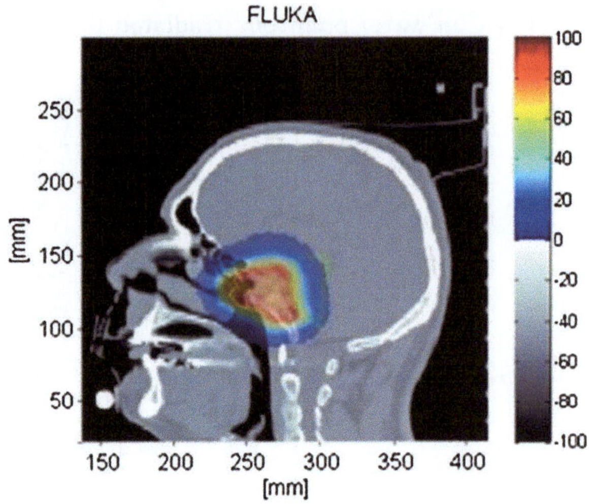

Fig. 2.29 Use of FLUKA to calculate the radiation dose distribution in a patient. The *color bar* shows the normalized values of the dose. *Courtesy* of Andrea Mairani (CNAO, Pavia, Italy)

verifying treatment plans. Till Böhlen, a researcher at the CERN Partner Project, is developing FLUKA for ion beam therapy. According to Böhlen, FLUKA is a valuable tool to accurately compute treatment doses, which is particularly useful in critical treatment care situations (e.g., the patient has a metal implant in the target area of intervention).

Future developments will include the development of improved FLUKA physical models for new ions, such as oxygen and helium, with a view to possible use in hadron therapy. The code is also widely used to simulate the secondary radiation that is produced during treatment when patient tissues interact with the beam. Secondary radiation is being studied as a very powerful tool to perform in vivo monitoring during treatment (Fig. 2.29).

In addition to FLUKA, GATE and GEANT4 software should be mentioned [28, 29]. GEANT4 has been used in applications involving particle physics, nuclear physics, accelerator design, space engineering, and medical physics. It was created by physicists and software engineers using object-oriented technology and is implemented in the C++ programming language. GEANT4 is a code used to simulate the passage of particles through matter, encompassing geometry, physical models and tips, which are very useful in electromagnetic, optical, and hadronic processes. It covers an energy range from 250 eV to TeV. GEANT4 has been widely used in simulations for hadron therapy.

A literature search to check the comparative effectiveness of FLUKA indicates that it is still difficult to give any recommendation. The problem should first be studied in more detail [30–32]. CERN comprehensively supports the development and use of FLUKA. It is also used at the Centro Nazionale di Adroterapia Oncologica, Pavia, Italy, where CERN researchers, mainly Entervision fellows, seek experimental data for their theses.

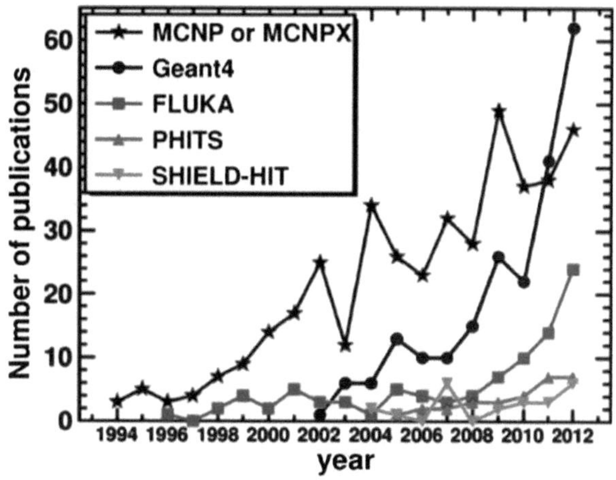

Fig. 2.30 Annual number of publications related to hadron therapy, where respective Monte Carlo codes/tools were used. *Source* Web of Science database (Thomson Reuters), October 2013. *Courtesy* Prof. Dr. Angela Bracco, NuPECC Chair

Figure 2.30 shows the number of publications related to hadron therapy by year, where different codes/tools in the Monte Carlo simulation are used. As shown, GEANT4 is predominately used today. However, FLUKA's flexibility and satisfactory agreement with dosimetry data and production of nuclear fragments indicate that the code is a valuable for supporting a wide variety of applications for proton and carbon ion therapy.

References

1. Nunes MA (2013) Large Hadron Collider: Nova era de descobertas. Hadronterapia. Seven System International Ltda, São Paulo
2. Amaldi U (2015) Particle accelerators: from big bang physics to hadron THERAPY. Springer, New York
3. http://commons.wikimedia.org/wiki/CERN
4. Alonso JR (2000) Review of ion beam therapy: present and future. In: Proceedings of EPAC, Vienna, Austria, p 235
5. DeLaney TF, Kooy HM (eds) (2008) Proton and charged particle radiotherapy. Lippincott Williama & Wilkins, Philadelphia
6. http://www.slac.stanford.edu/pubs/beamline/27/1/27-1-panofsky.pdf
7. Lincoln D (2009) The quantum frontier: the large hadron collider. The Johns Hopkins University Press, Baltimore
8. Amaldi U, Bonomi R, Braccini S et al (2010) Accelerators for hadrontherapy: from Lawrence cyclotrons to linacs. Nucl Instrum Methods Phys Res A 620:563–577
9. Chao AW, Chou W (eds) (2009) Reviews of accelerator science and technology. In: Medical applications of accelerators, vol 2. World Scientific, Singapore
10. Nunes MA (2013) Hadron therapy physics and simulations. Springer, New York
11. Amaldi U, Braccini S, Citterio A et al (2009) Cyclinacs: fast-cycling accelerators for hadrontherapy. http://arXiv:0902.3533(physics.med-ph)
12. http://en.wikipedia.org/wiki/Mark_Oliphant

13. Giudice GF (2010) A zeptospace odyssey: a journey into the physics of the LHC. Oxford University Press, Oxford
14. Trbojevic D et al (2011) Lattice design of a rapid cycling medical synchrotron for carbon/proton therapy. In: Proceedings of IPAC 2011, San Sebastian, Spain. http://www.teambest.com/news/BPT_CarbonIonSynchrotron_4page_Oct2014_MedPhys1.pdf
15. http://accelconf.web.cern.ch/accelconf/pac2013/papers/frxb1.pdf
16. Fourkal E, Shahine B, Ding M et al (2002) Particle in cell simulation of laser-accelerated proton beams for radiation therapy. Med Phys 29:2788–2798
17. Bulanov SV, Esirkepov TZ, Khoroshkov VS et al (2012) Oncological hadrontherapy with laser ion accelerators. Phys Lett A 299:240–247
18. Schippers JM, Lomx AJ (2011) Emerging technologies in proton therapy. Acta Oncol 50:838–850
19. Amaldi U (2007) Hadrontherapy: applications of accelerator technologies to cancer treatment. In: TERA Foundation Conference Presentation. http://basroc.rl.ac.uk/basroc_files/icpt/…/RAL-Amaldi-17.05.07.pdf. Accessed 17 May 2007
20. http://iopscience.iop.org/1748-0221/7/01/C01085/pdf/jinst12_01_c01085.pdf
21. Cerutti F, Battistoni G, Capezzali G et al (2006) Low energy nucleus–nucleus reactions: the BME approach and its interface with FLUKA. In: Proceedings of 11th international conference on nuclear reaction mechanism, Varenna, Italy, pp 507–511, 12–16 June. http://www.mi.infn.it/egadioli/Varenna2006/Proceedings/CeruttiF.pdf
22. http://sourceforge.net/projects/auflukatools/
23. http://www.fluka.org/content/manuals/FM.pdf
24. http://www.fluka.org/fluka.php?id=license
25. http://www.isgtw.org/spotlight/physics-software-used-fight-cancer
26. https://www.fluka.org/fluka.php?id=secured_intro
27. Andersen V, Ballarini F, Battistoni G et al (2004) The FLUKA code for space applications: recent developments. Adv Space Res 34:1302–1310
28. Agostinelli S, Allison J, Amako K et al (2003) Geant4—a simulation toolkit. Nucl Instrum Methods A 506:250–303
29. Allison J, Amako K, Apostolakis J et al (2006) Geant4 developments and applications. IEEE Trans Nucl Sci 53:270–278
30. Battistoni G, Muraro, S, Salaet PR et al (2007) The FLUKA code: description and benchmarking. In: Proceedings of the Hadronic Shower Simulation Workshop 2006: AIP Conference Proceedings, vol 896, pp 31–49
31. Battistoni G, Cerutti F, Engel R et al (2006) Recent developments in the FLUKA nuclear reaction models. In: Proceedings of 11th international conference on nuclear reaction mechanism, pp 483–495. Varenna, Italy, 12–16 June
32. Böhlen TT, Cerutti F, Dosanjh M et al (2010) Benchmarking nuclear models of FLUKA and GEANT4 for carbon ion therapy. Phys Med Biol 55:5833–5847
33. Jongen Y et al (2010) Proceedings of CYCLOTRONS. Lanzhou, China

Chapter 3
Physical and Biological Rationale for Using Ions in Therapy

3.1 Biophysical Properties

3.1.1 Stopping Power and Linear Energy Transfer

3.1.1.1 Stopping Power

In the physical processes of particle transport, the calculations take into account the following:

1. The loss of energy by ionization
2. Multiple Coulomb scattering
3. Nuclear fragmentation reactions

The theory of "stopping power" was fully developed by 1933. The important fact to note is that protons stop in solid or liquid media. Beyond the stopping point, the radiation dose is negligible. The range of the proton is proportional to the square of its kinetic energy. If the proton beam is mono-energetic, all protons stop at approximately the same depth. The speed at which the proton loses energy increases when it slows down; for a given proton–electron collision, more momentum is transferred to the electron (the proton either remains in or leaves the neighborhood). Thus, the stopping power depends on the energy and material that slows it down. When corrected for density, material such as lead ($Z = 82$) has less stopping power than materials such as beryllium ($Z = 4$), water, or plastics.

The term *stopping power* is commonly used to signify the average energy loss per unit path length and is measured in MeV/cm. The stopping power depends on the type and energy of the particle and the properties of the material. The ionization density along a path is proportional to the stopping power of the material. Accordingly, $S(E) = -\mathrm{d}E/\mathrm{d}x$, where S is the stopping power, E is energy, and x is

M. d'Ávila Nunes, *Protontherapy Versus Carbon Ion Therapy*,
Biological and Medical Physics, Biomedical Engineering,
DOI 10.1007/978-3-319-18983-3_3

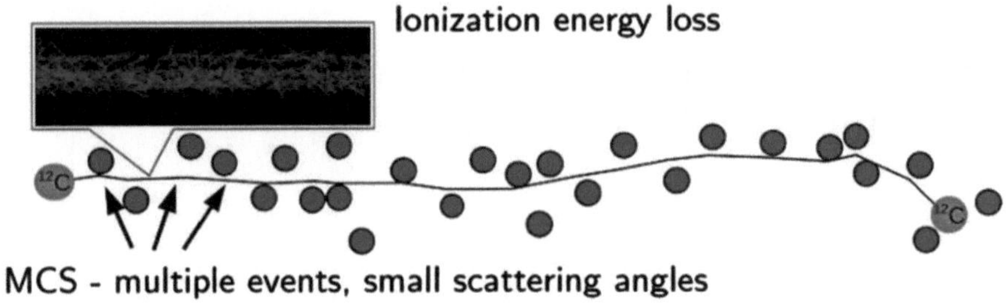

Fig. 3.1 Energy loss by ionization. Multiple Coulomb scattering (MCS) with multiple events. Reproduced from Nunes [1]

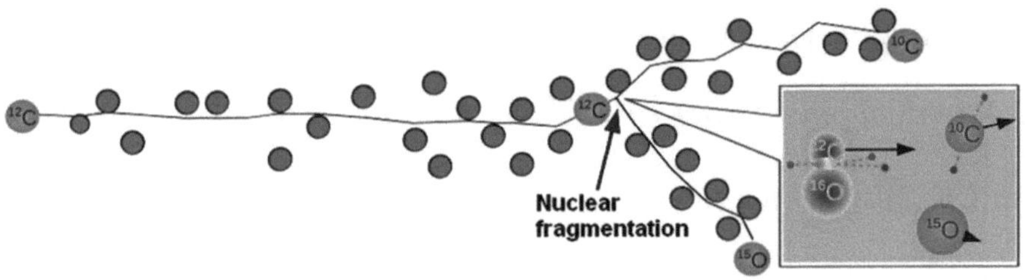

Fig. 3.2 Nuclear fragmentation. Reproduced from Nunes [1]

the path length. The minus sign makes S positive [2]. The stopping power increases toward the end of the ion range and reaches a maximum, the Bragg peak, immediately before the energy drops to 0. The curve describing this is termed the *Bragg curve*. The deposited energy can be derived by integrating the stopping power over the entire length of the path of the ion.

Scattering theory was first published in 1947. In general, the deflection of a proton by a single atomic nucleus is weak and the angular scattering observed regarding a beam of protons is due to a random combination of many of these deflections. Because of this and the electromagnetic interaction, scattering is more properly known as multiple Coulomb scattering (MCS; Fig. 3.1). The spatial distribution is approximately Gaussian. The MCS theory accurately predicts the width of this Gaussian distribution because the energy of the proton, the type of material, and thickness are known.

As mentioned, regarding the energy loss by ionization, $-dE/dx$ corresponds to $1/\beta^2$ and equals $1/E$ (continuous drop: formulas, tables, or approximations). With MCS, formulas and approximations are used. Nuclear fragmentation consists of simple events, and many secondary particles of several species are produced at large angles (this process is difficult to describe using a formula; see Fig. 3.2). A diagrammatic representation of a nucleus–nucleus collision is shown in Fig. 3.3.

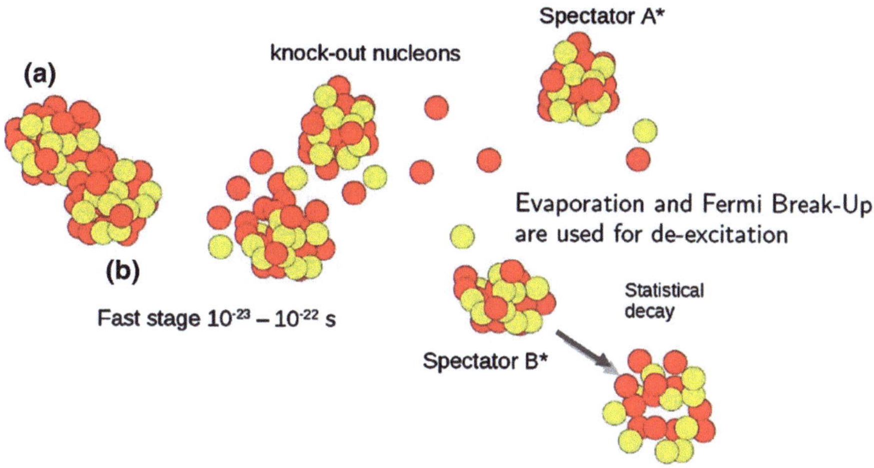

Fig. 3.3 Diagrammatic representation of a nucleus–nucleus collision. Reproduced from Nunes [1]

3.1.1.2 Linear Energy Transfer

The term *linear energy transfer* (LET) is often used in dosimetry and describes the effect of radiation on the matter. It represents the retarding force acting on ionizing charged particles traveling through matter—that is, how much energy ionizing particle material passed through the transfer unit of distance. LET depends on the nature of the radiation and the material traversed [3]. Thus, a high LET radiation will attenuate the radiation more rapidly, preventing deeper penetration. The high concentration of deposited energy can cause more severe injury to any microscopic structure near the particle track. LET is expressed as keV/μm or MeV/cm.

Despite the similarity between LET and stopping power, it must be remembered that they are different because stopping power has a nuclear component, and this component does not cause electron excitation. The carbon ions have sufficiently high LET components to provide biological efficiency as well as superior depth-dose distribution. The LET of carbon ion beams strongly increases in the incidence point on the body with increasing depth, reaching a maximum in the peak area (Fig. 3.4); in contrast, the LET beams of neutrons remains uniform at any depth in the body. This property is extremely advantageous from the therapeutic point of view in terms of its biological effect in the tumor. Thus, carbon ion beams form a large peak in the physical body due to an increase of dose and its biological efficacy to proceed at greater depths in the body. This fact is very important because of its potential use in the treatment of intractable cancers that are resistant to photon beams.

The linear energy transfer of charged particles in a medium is $L = dE/dl$, where dE is the average energy lost due to collisions and dl is the distance traversed by the particle. The behavior of increasing values of LET and relative biological effectiveness (RBE) are shown in Fig. 3.4.

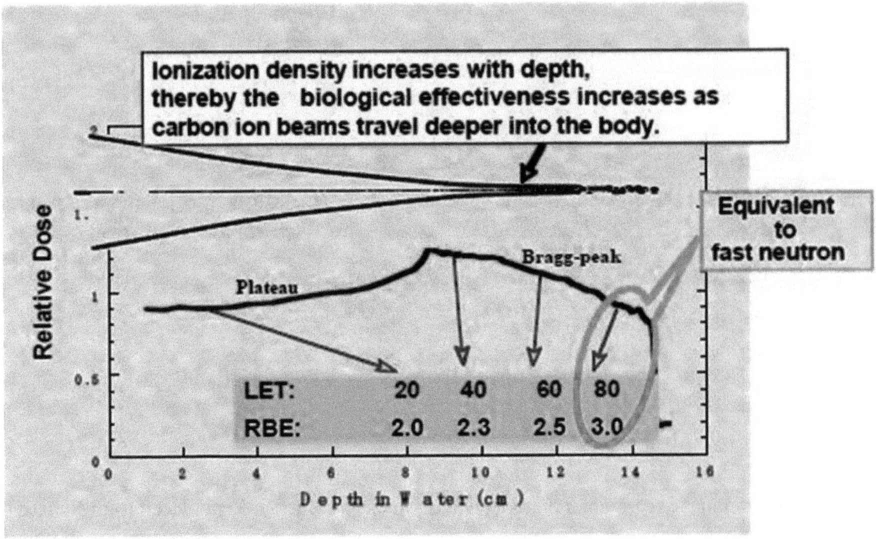

Fig. 3.4 Comparison of linear energy transfer (LET) and relative biological effectiveness (RBE) for carbon ions at 290 MeV. Spread-out Bragg peak = 60 mm. *Courtesy* National Institute of Radiological Sciences (NIRS)

3.1.2 *Radiation Dose*

3.1.2.1 Absorbed Dose

The absorbed dose is the energy deposited per unit mass in the medium (target) during exposure to ionizing radiation. It is measured in Gray (Gy), with 1 Gy = 1 J/kg. Thus, 1 Gray represents the absorption of 1 J of radiation per 1 kg of matter using the International System of Units. Because 1 J = 10^7 ergs and 1 kg = 1,000 g, 1 Gy = $10^7 \times 10^{-3}$ erg/g = 10^4 erg/g = 100 cGy.

3.1.2.2 Equivalent Dose

The equivalent dose takes into account the biological effectiveness of different types of radiation. It is calculated by multiplying the absorbed dose by a radiation weighting factor appropriate to the type and energy of the radiation and can be calculated using the following equation:

$$H_T = \Sigma W_R \times D_{T,R},$$

where H_T is the equivalent dose absorbed by tissue T, $D_{T,R}$ is the absorbed dose in tissue T from radiation type R, and W_R is the radiation weighting factor. H is measured in Sievert (Sv) using the International System of Units, named after the Swedish physician Rolf Maximilian Sievert (1896–1966) who studied the biological effects of radiation. The weighting factor is estimated using the relative biological effectiveness (RBE) value of a given radiation type at low doses and not high doses.

For high doses, the International Commission on Radiological Protections recommends a different definition of the equivalent dose:

$$H_e = \Sigma \mathrm{RBE}_{\mathrm{Re}} \times D_R,$$

where H_e is the equivalent dose and the weighting factors are replaced by the RBE value for a given radiation type (R) using a specific end point (e). The unit used here for dose is the Gray equivalent (GyE).

3.1.2.3 Effective Dose

The effective dose takes into consideration the dose due to ionizing radiation delivered nonuniformly. It takes into account both the type of radiation and the nature of each organ being irradiated. For a nonuniform radiation exposure, a different tissue weight factor (W_T) is used to reflect the different radiogenic sensitivity from different organs. The effective dose is given in Sv and is calculated as follows:

$$H_E = \Sigma W_T . H_T,$$

where H_E is the effective dose, W_T is the tissue weight factor defined by regulation, and H_T is the equivalent dose absorbed by tissue T.

3.1.2.4 Relative Biological Effectiveness

When high doses are employed in particle therapy, the GyE unit has been used for heavy ions, whereas researchers working with protontherapy still use the cobalt Gray equivalent (CGE). The International Commission on Radiation Units and Measurements proposed replacing this unit with RBE-weighted absorbed dose, defined as $D_{\mathrm{RBE},V} = \mathrm{RBE} \times D_V$. The unit is given in Gy (RBE). The volume (v) must be specified and may correspond to the tumor volume or the planned target volume.

3.1.2.5 Integral Dose

The integral dose is defined as the average dose deposited in the total irradiated volume of the patient, when comparing different radiation therapy types. It is the product of the mass of tissue irradiated and the absorbed dose, given by the following equation:

$$ID = mD$$

where ID is the integral dose, m is the mass, and D is the absorbed dose. The unit used is Kg × Gy.

3.1.2.6 Isoeffective Dose

The isoeffective dose D_{IsoE} is the dose of a radiation treatment carried out under reference conditions that produces the same clinical effects on the target volume as those of the actual treatment. It is the product of the total absorbed dose (in Gy) and a weighting factor W_{IsoE}, defined by the equation $D_{IsoE} = D \times W_{IsoE}$. The weighting factor takes into account all clinical factors that could influence the clinical effects: the dose per fraction, the total time, the quality of radiation, the biological system, and effects. In fractionated conventional X-ray radiation therapy, the dose per fraction and the overall treatment time are the two main parameters that can be adjusted. The weighting factor for an alteration of the dose per fraction is commonly evaluated using the linear-quadratic (α/β) model. For therapy with protons and heavier ions, radiation quality has to be taken into account. The isoeffective dose for heavy ion therapy is problematic because of the complex RBE field [4].

3.2 Biological Properties

3.2.1 Relative Biological Effectiveness

In June 2004 in Vienna, the first technical meeting was held to discuss the RBE of carbon ions compared to photons, with financial support obtained jointly from the International Atomic Energy Agency and International Commission on Radiation Units and Measurements [5]. Ion therapy, using carbon ions in particular, shows increased RBE compared with protons and photons (i.e., high LET vs. low LET). During this technical meeting, requests were made regarding the reporting of homogeneity. To assess the merits of ion therapy, it is essential that treatments are reported in a comparable way across all centers so that clinical reports and protocols can be interpreted without errors.

As previously discussed, the use of ion beams for radiation therapy was created at the Lawrence Berkeley National Laboratory at the University of California, and it is currently being investigated in Japan, Germany, and Italy. Ion beams have two important characteristics in the physical aspects of the dose distribution in the patient (i.e., high linear energy transfer), the path of the particle localized in tumor volume, and relative biological effectiveness. The most important of these biological phenomena is the marked efficiency in killing cells. The concept of RBE was introduced to account for this increased efficiency, defined as the ratio of the photon dose in relation to any other particle dose to produce the same biological effect. RBE cannot be defined specifically for a given radiation. This is a simple concept, but it has complex clinical implications because it is a function of the type of particle, energy, dose, dose per fraction, number of fractions, degree of oxygenation, tissue or cell type, and range from the early to late reactions following therapy.

Radiation therapy delivers a lethal dose to tumor cells while preserving healthy cells. Mortality is related to the damage induced in the DNA chain in the cell. The damage can be of two types:

1. Direct: The radiation breaks the chemical bonds of DNA
2. Indirect: Radiation-induced free radicals

Typically, these types of damage are repaired via a complex mechanism. Some of these mechanisms are impaired in tumor cells. Thus, different mortality occurs from exposure to the same radiation between tumor cells and normal cells. Different types of radiation have the same dose released to the target volume but a different effectiveness in killing tumor cells, as seen in Fig. 3.5 [6].

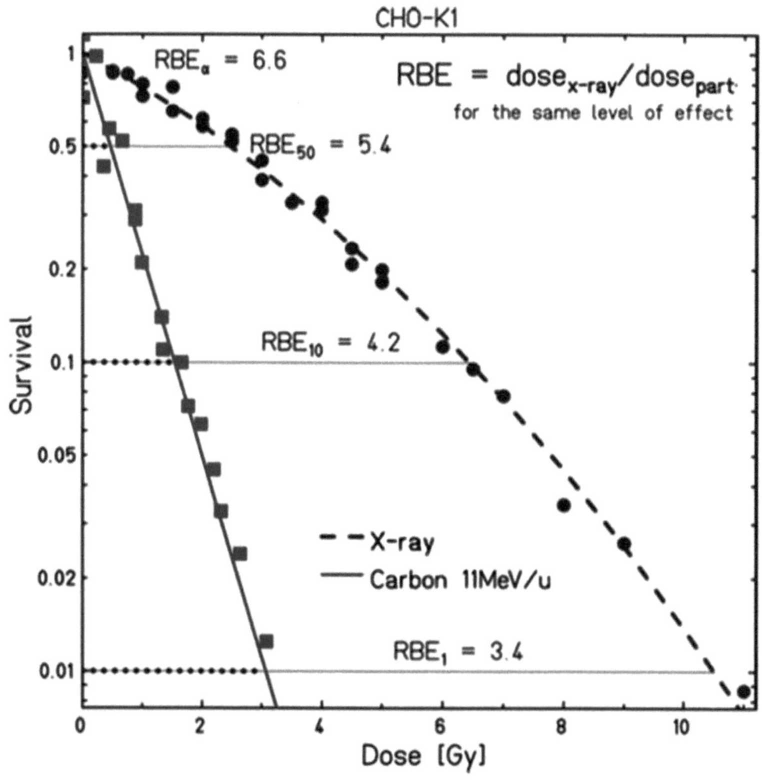

Fig. 3.5 Clonogenic survival curves illustrating the higher efficiency of carbon ions compared with X-rays. *Courtesy* Prof. Dr. Wilma K. Weyrather

RBE is defined as the ratio of the absorbed dose from a reference radiation, typically 1.2 MeV photons emitted by ^{60}Co, and the test radiation required to produce the same biological effect of a particular cell type.

$$\mathrm{REB} = \left.\frac{D_{\mathrm{Photon}}}{D_{\mathrm{Ion}}}\right|_{\mathrm{Isoeffect}}$$

The doses are usually taken at the level of 10 % of cell survival. For protons, the adopted RBE is equal to 1.1 on a clinical scale, meaning that protons produce the same lethal damage as the photon beams of conventional therapy. Hence, this is a simplified comparison for conventional therapy protocols and protontherapy, from a clinical point of view. However, for carbon ions, RBE is between 2 and 5, increasing in value in the last centimeters before the end for ion variation. This can be explained by the high ionization density caused by carbon ions passing through the cells—20 times greater than that produced by protons in the same range; as a consequence, it causes more destructive damage to DNA. The RBE thus depends on the type of particle and many other parameters, such as particle LET, dose, type of tissue, and tumor oxygenation state, among others. Importantly, at HIT, software development for treatment with carbon ions involved research conducted at GSI to obtain a radiobiological model (local effect model) to compute the RBE:

$$\mathrm{RBE\ (Proton)} \sim 1.1\ \mathrm{RBE\ (Carbon\ Ion)} \sim 2\ldots5$$

Consideration of dose-dependent and tissue-specific RBE values is crucial for hypofractionation studies, for example. An analysis of tumor control probability (TCP) for non-small cell lung cancer is an example of clinical results in terms of clinically determined RBE [7]. Figure 3.6 shows a method for determining RBE in

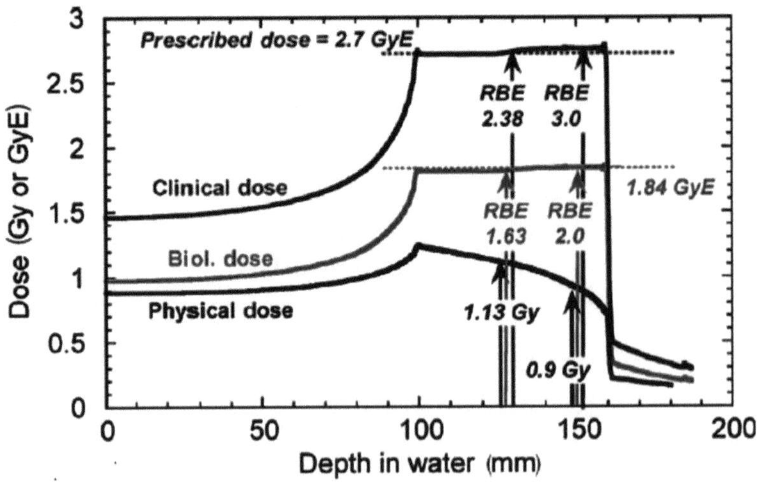

Fig. 3.6 Schematic method used to determine the RBE at the center of the spread-out Bragg peak for the clinical situation. *Courtesy* Emma Thornton, Academic Rights and Journals, Oxford University Press [11]

Table 3.1 Clinical RBE of the carbon beams at the center of the various sizes of spread-out Bragg peak

SOBP width (mm)	Clinical RBE
30	2.8
40	2.6
60	2.4
80	2.3
100	2.2
120	2.1

Courtesy Emma Thornton, Academic Rights and Journals, Oxford University Press [11]

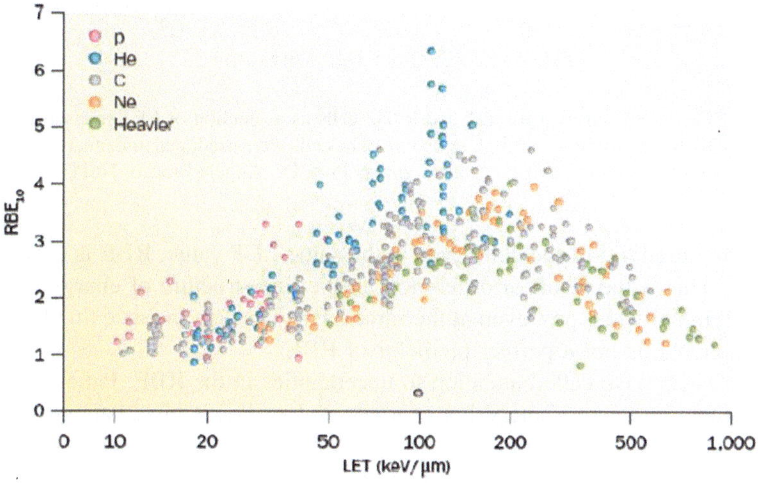

Fig. 3.7 RBE versus LET from published experiments on in vitro cell lines. RBE is calculated at 10 % survival; LET values are given as keV/μm in water. The *different colors* indicate different ions, from protons to heavy ions [8]. Data points are extracted from the Particle Radiation Data Ensemble (PIDE) database (www.gsi.de/bio-pide), which currently includes 855 survival curves for cells exposed to photons (alpha/beta ratio ranging from 1 to 30) and ions. *Courtesy* Prof. Dr. Angela Bracco, NuPECC Chair

the center of the spread-out Bragg peak (SOBP) for clinical situations. The clinical RBE of carbon ion beams varies with the size of the SOBP. The values measured in the center of the SOBP are presented in Table 3.1. Figure 3.7 shows the behavior of RBE versus LET for published experiments with cell lines in vitro.

Figure 3.8 shows the behavior of RBE of cell survival for T-1 and R2D2 cells as a function of LET 100–200 keV/μm (measured at various points in SOBPs for carbon, neon, and argon ions). A decline begins in the "overkill" region, where the amount of energy deposited in a cell by a single particle traversal is in excess of the

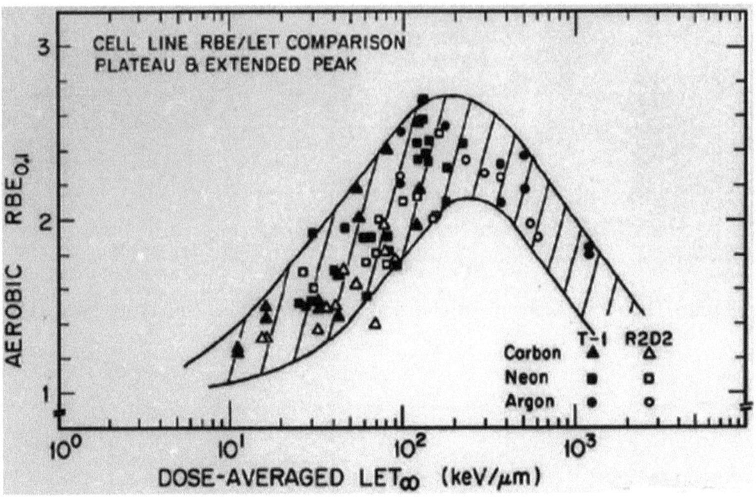

Fig. 3.8 RBE of cell survival for T-1 and R2D2 cells as a function of LET measured at various points in SOBPs for carbon, neon, and argon ions. The cells were irradiated under aerobic conditions and the end point was 10 % cell survival. *Courtesy* Prof. Dr. Angela Bracco, NuPECC Chair

amount required to kill the cell. Even at the same LET value, RBE is a function of ion type. This is the result of differences in the fine structure of energy deposition for different particle types, even at the same LET. The data indicate that LET, while often adequate, is not a perfect predictor of RBE.

Researchers have called attention to uncertainties in the RBE. Patients are being treated safely in several centers. For example, in Lanzhou, China, patients are treated for superficial and deep tumors with carbon ions without any correction for RBE, using a flat SOBP in physical dose. Thus, the need is not for more RBE measurements, but rather for a newly emerging radiobiology with new scenarios in hadron therapy.

3.2.2 Oxygen Enhancement Ratio

RBE also depends on the oxygen content in tumor cells. Oxygen is known as a radiosensitizer, which plays an important role in radiotherapy because tumors with low vascularization may become radioresistant [9]. We can quantify this by applying the following equation:

$$\mathrm{OER}_{\mathrm{particle}} = D_{\mathrm{anoxic}}/D_{\mathrm{oxic}}\big|_{\mathrm{isoeffect}}$$

The oxygen enhancement ratio (OER) decreases as the LET increases. The OER for carbon ions is close to 1 and can reach as high as 3 to photons and protons. This can be explained by the difference in DNA damage induced by carbon ions and photons.

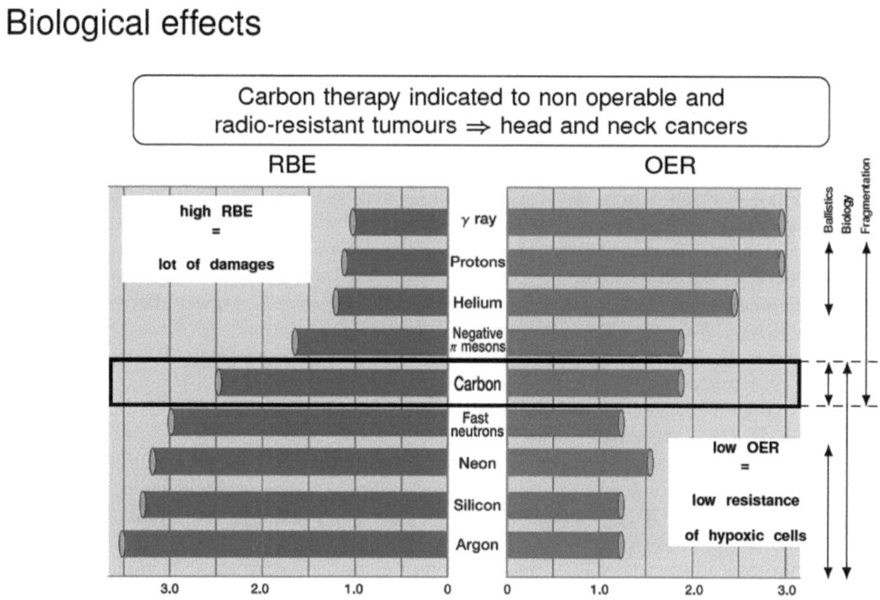

Fig. 3.9 The behavior of RBE and OER as a function of radiation employed. *Courtesy* Prof. Dr. Daniel Cussol

This is important because it extends the range of tumor indication that can be successfully treated by radiation, thus strongly supporting carbon ion therapy over protontherapy.

Figure 3.9 shows the behavior of RBE and OER based on the radiation used. The increased rate of oxygen (OER) refers to the increase in the therapeutic effect of ionizing radiation due to the presence of oxygen. The increased amount of oxygen creates free radicals and increases the damage to the target tissue. In solid tumors, the innermost regions can have much less oxygen than normal tissue, and triple the radiation dose is required to achieve the same probability, as in the case of the control in tumors with normal oxygenation.

The dependence of RBE and OER in LET was studied by Brendsen in the early 1960s. He showed that the RBE reaches a maximum at an LET of 100–200 keV/μm —the same LET in which the OER decreased to approximately 1.0. More recent studies have shown that RBE has a peak at an LET value that is dependent on the particle, indicating that the LET alone does not adequately define energy deposition and its influence on biological effect. One of the complications associated with heavy ion beams and pions is the increase of RBE with depth at the stop region. With heavy ions, the RBE is also dependent on dose and the dose fractionation scheme used.

References

1. Nunes MA (2013) Hadron therapy physics and simulations. Springer, New York
2. Berger MJ, Coursey JS, Zucker MA, Chang J (1998) Stopping power and range tables for electrons, protons, and helium ions. NIST Standard Reference Database 124. http://www.nist.gov/pml/data/star/
3. International Commission on Radiation Units and Measurements (1970) Linear energy transfer. ICRU Report 16
4. Newhauser WD, Durante M (2011) Assessing the risk of second malignancies after modern radiotherapy. Nat Rev Cancer 11:438–448
5. International Atomic Energy Agency (IAEA) (2008) Relative biological effectiveness in ion beam therapy. Technical Reports Series No. 461. Vienna
6. Mihailescu D, Borcia C (2012) Biophysical models in hadrontherapy. J. Adv. Res. Phys. 3(1):1–9. http://stoner.phys.uaic.ro/jarp/index.php/jarp/article/download/105/62
7. International Commission on Radiation Units and Measurements (1970) Linear energy transfer. ICRU, Report 16
8. NuPECC (2013) Nuclear physics in medicine. hadrontherapy, Chapter I. European Science Foundation. http://www.esf.org/uploads/media/Nuclear_Physics_in_Medicine.pdf
9. Timlin C, Jones B (2010) Proton and charged particle radiotherapy. Br J Radiol 83:87
10. http://www.lpc-caen.in2p3.fr/root/.EJC2011-CUSSOL/EJC2011_CUSSOL.pdf
11. Naruhiro Matsufuji et al. (2007) Specification of carbon ion dose at the National Institute of Radiological Sciences (NIRS). J Radiat Res 48 (Suppl A): A81–A86 doi:10.1269/jrr.48.A81

Chapter 4
Modelling Heavy Ion Radiation Effects

4.1 Biophysical Models

When planning treatment, RBE values have to be estimated as precisely as possible. The RBE has complex dependencies, so it is important to simplify processes in order to identify and quantitatively describe the most important when applying biophysical models in ion beam therapy (IBT). The primary particles and also all fragments produced in the stopping process should be considered because the biological effect depends on the particle [1]. The biological response to radiation by ions also depends on the particle energy, dose level, oxygen status, and the irradiated tissue or cell system [2]. As noted, all of these factors must be considered to predict the response of biological tissue to the complex radiation field.

The facilities that treat cancer patients with carbon ions use different strategies (Fig. 4.1) [3]. At HIMAC (Chiba, Japan), an experimental approach based on precise measurements of RBE in vitro is being used (Fig. 4.2); these measurements are used to determine the isoeffective shape of the profile depth dose. Clinical values of RBE are determined by a combination of clinical experience with neutrons, which show radiobiological characteristics similar to beams of carbon ions at the end of the depth of penetration. Therefore, the clinically determined RBE for a neutron beam was 3.0 when the total number of divided doses was 18 and the neutron dose level for each fraction was 0.9 Gy. The clinical value of RBE to an equivalent position to neutrons of SOBP for carbon ions was then determined as 3.0. For protons, an RBE value of 1.1 is used.

At GSI, a local effect model is used (Fig. 4.3) [3].

Next, a direct comparison of protons and carbon ions is shown, analyzing survival (in log scale) for Chinese Hamster Ovary (CHO) cells, depending on the depth. CHO cells are epithelial cells that grow adherent monolayers in culture; they are a hugely popular research tool in the molecular biology community. This is the first radiological experiment developed at HIT using protons and carbon ions and the corresponding models (Fig. 4.4).

M. d'Ávila Nunes, *Protontherapy Versus Carbon Ion Therapy*,
Biological and Medical Physics, Biomedical Engineering,
DOI 10.1007/978-3-319-18983-3_4

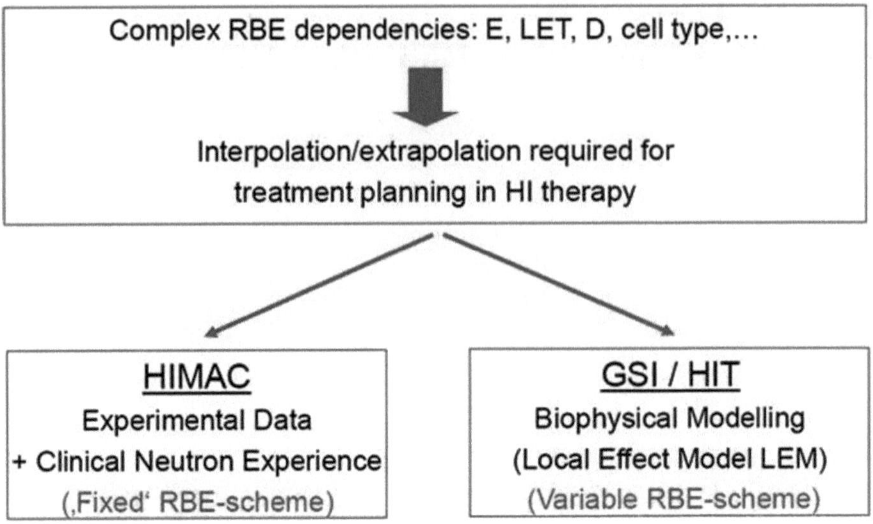

Fig. 4.1 Treatment planning for carbon ions. Reproduced from: Scholz [3]. *Courtesy* Prof. Dr. Thomas Friedrich on behalf of Prof. Dr. M. Scholz

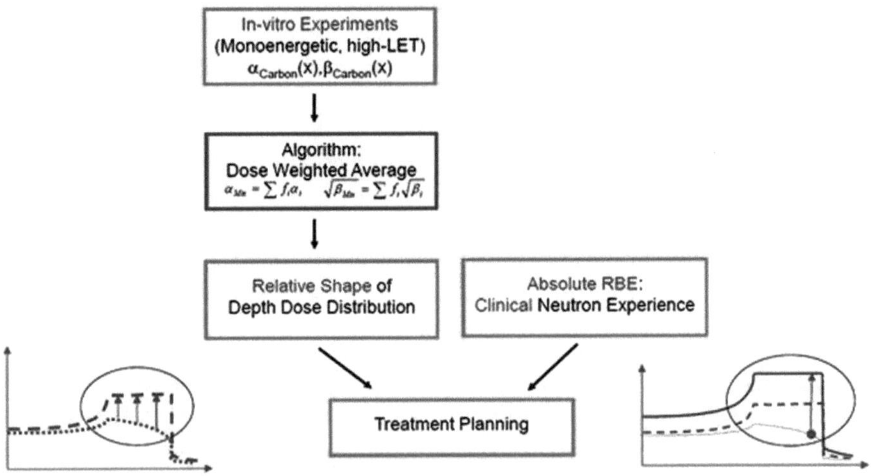

Fig. 4.2 HIMAC approach. Reproduced from: Scholz [3]. *Courtesy* Prof. Dr. Thomas Friedrich on behalf of Prof. Dr. M. Scholz

A comparison between the NIRS and GSI data shows a 15 % difference in the clinical dose in the middle of the SOBP. It is indispensable to establish conversion between GSI and other centers to make clinical experiences referenced and help to find an optimal treatment protocol using heavy ions, since the difference in results can be as great as 15 %

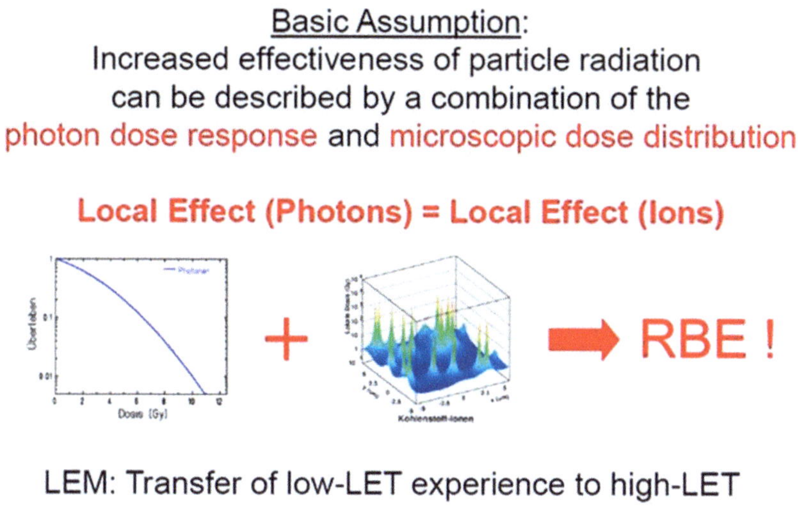

Fig. 4.3 GSI approach using a local effect model (LEM). Reproduced from: Scholz [3]. Courtesy: Prof. Dr. Thomas Friedrich on behalf of Prof. Dr. M. Scholz

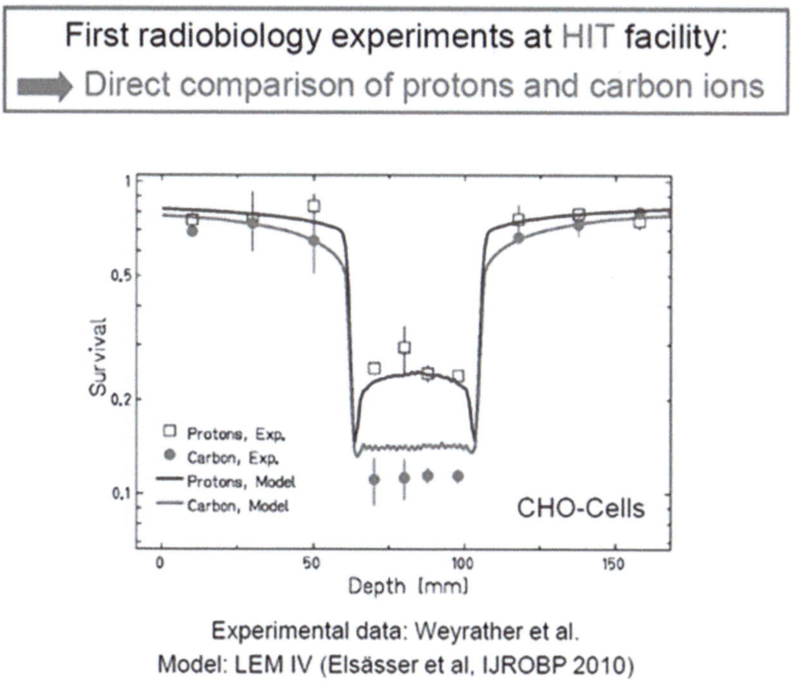

Fig. 4.4 Carbon ions versus protons. The protons (*solid blue line*) and carbon ions (*solid red line*) obtained using the model are in good agreement with the experimental data for CHO cells. *Courtesy* Elsevier and Copyright Clearance Center [11]

4.2 The Alpha/Beta Ratio

Various mathematical models of varying degrees of complexity have been developed to define the shape of the curves for cell survival. All of the models are based on the concept of random nature deposition of energy by radiation.

The linear-quadratic model is used to describe the curve of cell survival, assuming that there are two components of cell death by radiation:

$$S(D) = e^{-\alpha D - \beta D^2}$$

where *S(D)* is the fraction of surviving cells at dose *D*, α is a constant describing the initial slope of the cell survival curve, and β is a smaller constant describing the quadratic component of cell killing. The ratio of alpha to beta gives the dose at which the linear and quadratic components of cell killing are equal.

Although it has several limitations, this ratio is used in predicting clinical effects in response to radiation as one of parameters to model cell death by radiation. In radiotherapy (RT), the sensitivity to changes in fractionation can be quantified in terms of the alpha/beta ratio. For many human tumors, the ratio is high (typically 10 Gy). This ratio is obtained from isoeffect curves plotted using the survival fractions of a single cell line at different doses per fraction [4]. It is the byproduct of the linear quadratic model, which describes cell killing as a single-hit versus double-hit hypothesis: linear cell kill is expressed by the alpha component, whereas quadratic cell kill is expressed by the beta component. A high alpha/beta ratio (6–14 Gy), seen in many human tumors, suggests a predominance of alpha component, implying a decreased response to fractionation and, thus, a decreased clinical benefit of hyperfractioning. A low alpha/beta ratio (1.5–5 Gy) is usually associated with a delayed response of normal tissue and is the basis for the therapeutic gain achieved by using hypofractionation (Table 4.1).

Examples of tumors with low alpha/beta ratios are prostate cancer (2.7 Gy), rhabdomyosarcoma (0.4 Gy), and melanoma (0.6 Gy). These tumors theoretically benefit from treatment with carbon ions for two reasons: increased cell death beyond that achieved with RT photons as a result of a higher RBE and decreasing toxicity to normal tissue due to the depth of dose distribution with the carbon ions. Thus, it is necessary to use the alpha/beta ratio as a guide in selecting tumors. An alternative to using the alpha/beta ratio is to observe the survival fraction at 2 Gy of various tumors, as a substitute for radiosensitivity tumors irradiated by photons.

The alpha/beta ratio requires additional study despite the classification of tumors into five categories by Deacon et al. [5]. It is very difficult to measure the alpha/beta ratio in tumors. Furthermore, within the realm of personalized medicine, tumors that respond well to traditional treatment should continue to be treated with these therapies, reserving treatment with carbon ion therapy for those rare malignancies that do not provide adequate response. This treatment should be provided to the patient in a specific way—that is, control should be exercised to ensure that a patient is not treated outside of established protocols, thus achieving a breakthrough in the field and improvement in the patient's outcome.

Table 4.1 Alpha/beta ratios for normal human tissues and tumors [6]

Tissue/organ	End point	α/β ratio [Gy]	95 % conf. lim. [Gy]	References
Early reactions				
Skin	Erythema	8.8	[6.9;11.6]	Turesson and Thames (1989)
	Erythema	12.3	[1.8;22.8]	Bentzen et al. (1988)
	Desquamation	11.2	[8.5;17.6]	Turesson and Thames (1989)
Oral mucosa	Mucositis	9.3	[5.8;17.9]	Denham et al. (1995)
	Mucositis	15	[−15;45]	Rezvani et al. (1991)
	Mucositis	~8	?	Chogule and Supe (1993)
Late reactions				
Skin/vasculature	Telangiectasia	2.8	[1.7;3.8]	Turesson and Thames (1989)
	Telangiectasia	2.6	[2.2;3.3]	Bentzsn et al. (1990)
	Telangiectasia	2.8	[−0.1;8.1]	Bentzen and Overgaard (1991)
Subcutis	Fibrosis	1.7	[0.6;2.6]	Bentzen and Overgaard (1991)
Muscle/vasculature/cartilage	Impaired shoulder movement	3.5	[0.7;6.2]	Bentzen et al. (1989)
Nerve	Brachial plexopathy	<3.5	?	Olsen et al. (1990)
	Brachial plexopathy	~2	?	Powell et al. (1990)
	Optic neuropathy	1.6	[−7;6.2]	Jiang et al. (1994)
Spinal cord	Myelopathy	<3.3	?	Dische et al. (1981)
Eye	Corneal injury	2.9	[−4;10]	Jiang et al. (1994)
Bowel	Stricture/perforation	3.9	[2.5;5.3]	Deore et al. (1993)
	Various late effects	4.3	[2.2;9.6]	Dische et al. (1999)
Lung	Pneumonitis	4.0	[2.2;5.8]	Bentzen et al. (2000)
	Lung fibrosis [radiological]	3.1	[−0.2;8.5]	Dubray et al. (1995)
Head and neck	Various late effects	3.5	[1.1:5.9]	Rezvani et al. (1991)
	Various late effects	4.0	[3.3;5.0]	Stuschke and Thames (1999)
Supraglottic larynx	Various late effects	3.8	[0.8;14]	Maciejewski et al. (1986)
Oral cavity + oroph.	Various late effects	0.8	[−0.6;2.5]	Maciejewski et al. (1990)
Tumour				
Head and neck				
Larynx		14.5	[4.9;24]	Rezvani et al. (1993)
Vocal cord		~13	Wide	Robertson et al. (1993)

(continued)

Table 4.1 (continued)

Tissue/organ	End point	α/β ratio [Gy]	95 % conf. lim. [Gy]	References
Oropharynx		~16	?	Horiot et al. (1992)
Buccal mucosa		6.6	[2.9;infinity]	Maciejewski et al. (1989)
Tonsil		7.2	[3.6;infinity]	Maciejewski et al. (1989)
Nasopharynx		16	[−11;43]	Lee et al. (1995)
Various		10.5	[6.5;29]	Stuschke and Thames (1999)
Skin		8.5	[4.5;11.3]	Trott et al. (1984)
Prostate		?		Brenner and Hall (1999)
Melanoma		0.6	[−1.1;2.5]	Bentzen et al. (1989)
Liposarcoma		0.4	[−1.4;5.4]	Thames and Suit (1986)

Reproduced from reference: http://www.gphysics.net/index.php/tables.html?id=124

4.3 Local Effect Model

Theoretical models of cell inactivation induced by radiation include the target model, molecular (or linear-quadratic) model, theory of dual radiation action (TDRA), Katz's amorphous track structure model, local effect model (LEM), and microdosimetric kinetic model (MKM) [7]. Of these models, the most successful for carbon ion therapy are the LEM and MKM (Fig. 4.5). Models are not easily accessible from the experimental data (in vitro), indicating highly complex problems to be solved. However, LEM has advantages over other models and has been successfully used in treatment planning for carbon ion therapy. Using LEM, it is possible to transfer the results of in vitro studies to clinical cases. This model was proposed by Scholz and Kraft from GSI, in order to include RBE values in treatment planning for patients undergoing ion therapy; it has also been used together with MKM at Istituto Nazionale di Fisica Nucleare (Turin, Italy).

In LEM, it is mainly assumed that the biological effect is determined by the local dose, being independent of the particular type of radiation that leads to local power deposition. This concept is applied in volumes on the nanometer scale, not microns as in microdosimetry. The three main assumptions of the LEM are as follows:

1. *Photon survival curve* The linear quadratic model (alpha, beta) is used for parameterization of the dose-effect curve after irradiation by photon.
2. *Radial dose profile* The dose is supposed constant around a center of 10 nm, decreasing as the inverse square of the distance.
3. *Target geometry* The cell nucleus is admitted as cylindrical.

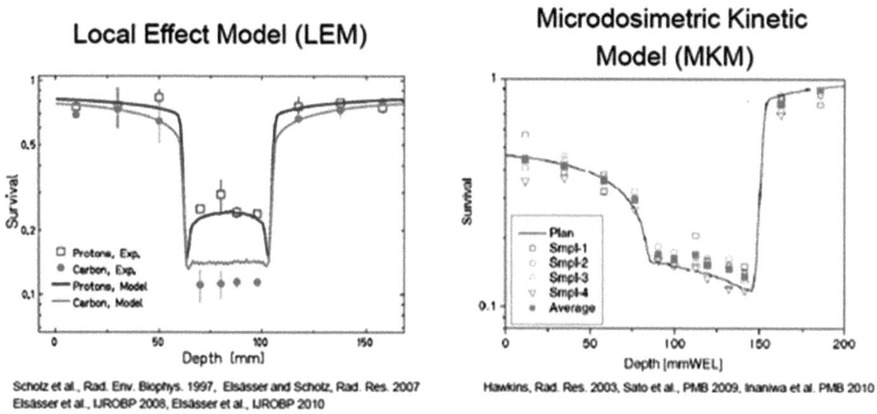

Fig. 4.5 Models for the prediction of RBE. *Courtesy* Elsevier and Copyright Clearance Center [12]

By these assumptions, one can obtain all required parameters and equations to work with the LEM. The LEM has been much criticized, mainly by assuming that equal numbers of events of local deposit imply that a low-LET survival curve can be used to achieve the effect produced by high-LET radiation. Clustered DNA damage was incorporated into the LEM, which significantly increased the accuracy of the predictions [8].

There are four versions of the LEM:

- LEM I: The original version proposed by Scholz and Kraft in1994, which also includes a version in which the approximate values of beta may be readily calculated from alpha [9]
- LEM II: Includes clustered damage effect in the DNA [10]
- LEM III: Includes an ion energy-dependent value to compensate for the systematic deviation in RBE predictions [11]
- LEM IV: Includes the relationship of the distribution of double-strand breaks (which is believed to be primarily responsible for cell death) in the characteristic volume [12]

Using the LEM IV [12] on for the measurement of ion beam radiotherapy for RBE, Elsasser et al. conducted a direct, experimental comparison between proton and carbon ion beams and a novel approach for treatment planning . They used CHO-K1 cells exposed in the three-dimensional phantom with the pencil beam scanning technique, and they compared the experimental data with a novel biophysical model. This new approach constitutes a more sophisticated consideration of spatially correlated damage induced by ion irradiation. The experimental data with the new approach show the advantages of carbon ions compared with protons for treatment-like field configurations. The model predicts the effectiveness for various ionic species with similar precision, making it a powerful tool for the further optimization and use of ion beams in tumor therapy. In addition, the predictions of the new biophysical model are clearly supportive of the potential benefits of carbon ions relative to protons for treatment, thus rebutting some criticism in the literature [13].

References

1. Schardt D, Elsässer T, Schulz-Ertner D (2010) Heavy-ion tumor therapy: physical and radiobiological benefits. Rev Mod Phys 82:383–425
2. Scholz M (2003) Effects of ion radiation on cells and tissues. Adv Polym Sci 62:96–155
3. Scholz M (2014) Radiation quality in ion beam therapy: how to take into account the RBE? EURADOS Winter School 2014. GSI Darmstadt
4. Schlaff CD, Krauze A, Belard A et al (2014) Bringing the heavy: carbon ion therapy in the radiobiological and clinical context. Radiat Oncol 9:88–107
5. Deacon J, Peckham MJ, Steel GG (1984) The radio responsiveness of human tumours and the initial slope of the cell survival curve. Radiother Oncol 2:317–323
6. http://www.gphysics.net/index.php/tables.html?id=124
7. Inaniwa T, Furukawa T, Kase Y et al (2010) Treatment planning for a scanned carbon beam with a modified microdosimetric kinetic model. Phys Med Biol 55:6721–6737
8. Mihailescu D, Borcia C (2012) Biophysical models in hadrontherapy. J. Adv. Res. Phys. 3(1):1–9. http://stoner.phys.uaic.ro/jarp/index.php/jarp/article/download/105/62
9. Scholz M, Kellerer AM, Kraft-Weyrather W, Kraft G (1997) Computation of cell survival in heavy ion beams for therapy. The model and its approximation. Radiat Environ Biophys 36:59–66
10. Elsasser T, Scholz M (2006) Improvement of the local effect model (LEM)—implications of clustered DNA damage. Radiat Prot Dosimetry 122:475–477
11. Elsasser T, Kramer M, Sholz M (2008) Accuracy of the local effect model for the prediction of biological effect of carbon ions beams in vitro and in vivo. Int J Radiat Oncol Biol Phys 71:866–872
12. Elsässer T, Weirather WK, Friedrich T et al (2010) Quantification of the relative biological effectiveness for ion beam radiotherapy: direct experimental comparison of proton and carbon ion beams and a novel approach for treatment planning. Int J Radiat Oncol Biol Phys 78(4):1177–1183
13. Wilkens JJ, Oelfke U (2008) Direct comparison of biologically optimized spread-out Bragg peaks for protons and carbon ions. Int J Radiat Oncol Biol Phys 709:262–266

Chapter 5
Clinical Experiences with Carbon Ion Therapy

5.1 Carbon Ion Therapy Facilities

Ion beam radiotherapy began with the use of proton beams at the Lawrence Berkeley National Laboratory (LBNL) in 1954. There, it was used heavy loaded cores as helium, carbon, nitrogen, neon, silicon, and argon, for clinical use, by measuring their effectiveness. In the period from 1977 to 1992, patients were treated with helium and neon ions. Japanese researchers interning at LBNL took the technique to Chiba, Japan, and initiated clinical studies with carbon ions using the HIMAC.

Currently, there are more than 30 protontherapy centers in operation, especially in the United States; carbon ion therapy is conducted in five facilities in the world, starting with the HIMAC/NIRS in Japan, followed by the GSI in Darmstadt in Germany (1997) and the HIT in 2009, where protons and carbon ions are used for clinical use. In 2001, the Hyogo Ion Beam Medical Center (HIBMC) was created in order to devote itself to the proton and carbon ions. Since 2006, the Institute of Modern Physics (IMP) in Lanzhou has developed clinical trials employing carbon ions for the treatment of superficial tumors with a system of only 100 MeV/u. In 2010, the Gunma University Heavy Ion Medical Center (GHMC) in Japan started clinical studies. In 2011, the Centro Nazionale di Adroterapia Oncologica (CNAO) was created in Italy, providing treatment with protons and carbon ions employing an excellent synchrotron to accelerate carbon ions. In Germany, in Marburg and Kiel, systems that are similar to the HIT are under construction. One was also built in Shanghai, China. Six centers—two in Japan, one in Austria, one in China, and two in the United States—are funded and in progress.

M. d'Ávila Nunes, *Protontherapy Versus Carbon Ion Therapy*,
Biological and Medical Physics, Biomedical Engineering,
DOI 10.1007/978-3-319-18983-3_5

5.2 What Are the Characteristics of Carbon Ions?

5.2.1 Physical Aspects

Carbon ions penetrate the tissue and deposit energy near the end of their variation, known as the Bragg peak. The peak is narrow, with the dose at the peak being much greater than the dose at the plateau region. The range of the particle is determined by the energy of incoming particles. Therefore, carbon ions are completely different from photons and neutrons, which have an exponential absorption dose versus depth.

The quality of dose distribution depends on the energy spread and range straggling, being smaller in magnitude when compared to protons. Also, the degree of lateral sharpness, called shadows, depends on the Coulomb scattering and becomes smaller with increasing particle mass [1]. Therefore, a lateral fall-off side around the target volume is faster for a beam of carbon ions compared to proton beams. In the distal end of the peak, almost no dose is deposited by protons, carbon ions but for a small amount due to the primary carbon ions undergo interactions and nuclear fragmentation at lower atomic number particles, producing a cluster beyond the peak tail. Certainly, the biological effect of this fragmentation is small because the tail contains only fragments with low atomic numbers.

Because the original peak is narrow and sharp, it has to be extended to conform to the shape and size of the lesions, which can be done using a beam scattering method with a passive system or a beam scanning method with an active beam release. In the passive method, the narrow peaks sweep a large area by a peak filter (ridge) in order to create a spreading of the Bragg peak (SOBP), corresponding to the shape and size of the target volume. In this method, a combination of a modulator, collimator, and compensator is used. In the active method, the peak position within the target is moved by a change in beam energy at the throttle or through the use of absorbers, precisely conforming the dose to the target volume.

5.2.2 Radiobiological Aspects

LET is defined as linear energy transfer or restricted linear collision stopping power. L_Δ) of charged particles in a medium is the quotient of *dE* by *dl*, where *dl* is the distance traversed by the particle and *dE* is the mean energy-loss due to collisions with energy transfers less than some specified value Δ.: $L_\Delta = (dE/dl)$. Photons, electrons and protons have low LET radiation. Neutrons and carbon ions have radiation with high LET. As discussed previously, RBE increases with increasing LET along the same SOBP. However, a LET range for which this is valid must be specified (10–120 keV/μm, in various experimental conditions).

Neutrons have uniformly high LET at any depth. For carbon ions, LET increases steadily with increasing depth, reaching a maximum in the peak region. This is an advantage from a therapeutic standpoint because the RBE of carbon ions increases with increasing depth for the tumor region. When different ionic species are compared, it is observed that the carbon ions show the biggest RBE ratio (peak/plateau). Thus, they have the best balance in terms of physical dose distribution and biological effect. Therefore, we can infer from this data that the use of carbon ions is highly effective in deeply situated tumors and is resistant to the photon beam.

Tumors with a low response to low LET radiation have a high proportion of hypoxic cells, poor oxygenation, and high capacity of repair. Thus, the benefit of high LET radiation is because the reduction in the OER is achieved by increasing the LET, along with the reduction in radiosensitivity-related differences in the position of the cells in the cell cycle.

5.3 How Is Treatment Planned?

The following steps make up the treatment planning process:

1. Systems for the immobilization of the patient are developed for each particular patient.
2. CT scans are consulted.
3. To determine the target volume, the CT image of mergers, magnetic resonance imaging, and positron emission tomography are used.
4. In the case of moving organs, the *gating technique* (described in [2]) is processed at the time of CT scans.
5. Treatment planning at HIT uses a GSI approach to select RBE, which is based on a local effect model. This model allows the calculation of RBE, based on the physical characterization of radiation fields and biological characterization of the response of cells or tissues, using the linear-quadratic approach.
6. Treatment planning at NIRS uses a model based on clinical experience, with high-LET neutron beams. To obtain the SOBP, a tumor cell line from the human salivary gland was selected as an in vitro model, due to the use by patients with adenocarcinomas.
7. Treatment planning at HIMAC uses the Hitachi Integrated Planning.
8. Treatment planning at CNAO uses two models: the local effect model and the microdosimetric kinetic model.

Globally, the LEM currently has greater acceptance than the other models. The difference between the results obtained at HIT and NIRS reach 15 %, which is significant. Corrections should be made when performing a comparison between the results obtained from both techniques.

Table 5.1 provides the indications for various types of cancer and the type of radiation employed by institutions.

Table 5.1 Indications for various types of cancer and the results obtained using photons and carbon ions at HIMAC-NIRS (Japan) and carbon ions at GSI (Germany) [3]

Indication	End point	Results photons	Results carbon HIMAC-NIRS	Results carbon GSI
Chordoma	local control rate	30 – 50 %	65 %	70 %
Chondrosarcoma	local control rate	33 %	88 %	89 %
Nasopharynx carcinoma	5 year survival	40 -50 %	63 %	
Glioblastoma	av. survival time	12 months	16 months	
Choroid melanoma	local control rate	95 %	96 % (*)	
Paranasal sinuses tumours	local control rate	21 %	63 %	
Pancreatic carcinoma	av. survival time	6.5 months	7.8 months	
Liver tumours	5 year survival	23 %	100 %	
Salivary gland tumours	local control rate	24-28 %	61 %	77 %
Soft-tissue carcinoma	5 year survival	31 – 75 %	52 -83 %	

Similar to protons

Table by G. Kraft 2007

Results of carbon ions

Geneva - 16.10.13 - UA

Courtesy Prof. Dr. Gerhard Kraft

5.4 Carbon Ion Exploration in Future Clinical Trials

Currently, clinical trial design is based on the assumption that the same biological effective dose is administered by photons, protons, or carbon ions. However, it is necessary to equate these doses, breaking the use of the traditional reference to photon dose. Therefore, it is important to look at differences in RBE and tissue type in order to create the best therapeutic ratio.

The design of future clinical screenings should address the differences in radiosensitivity between radio-resistant cells for low-LET RT and sensitivity for carbon ion RT to enable proper selection of histologies and patients who may benefit from this modality based on biomarkers and imaging. Patient selection should follow the protocol outlined in figure [4], acknowledging that none of the individual measures, whether hypoxia, alpha/beta ratio, or tumor proliferation, may represent the tumor microenvironment and true radiosensitivity of the tumor.

In Fig. 5.1, a cumulative score greater than or equal to 5 describes a radio-resistant tumor for low LET irradiation with standard of care (SOC), causing significant toxicity. Tumor histology and patients with 0–1 score have significant benefits from SOC treatment. The use of carbon ion therapy is sensitive when the advantages of using carbon ions exceed the therapeutic advantages that can be obtained with fractionated photon RT. With the advent of personalized medicine,

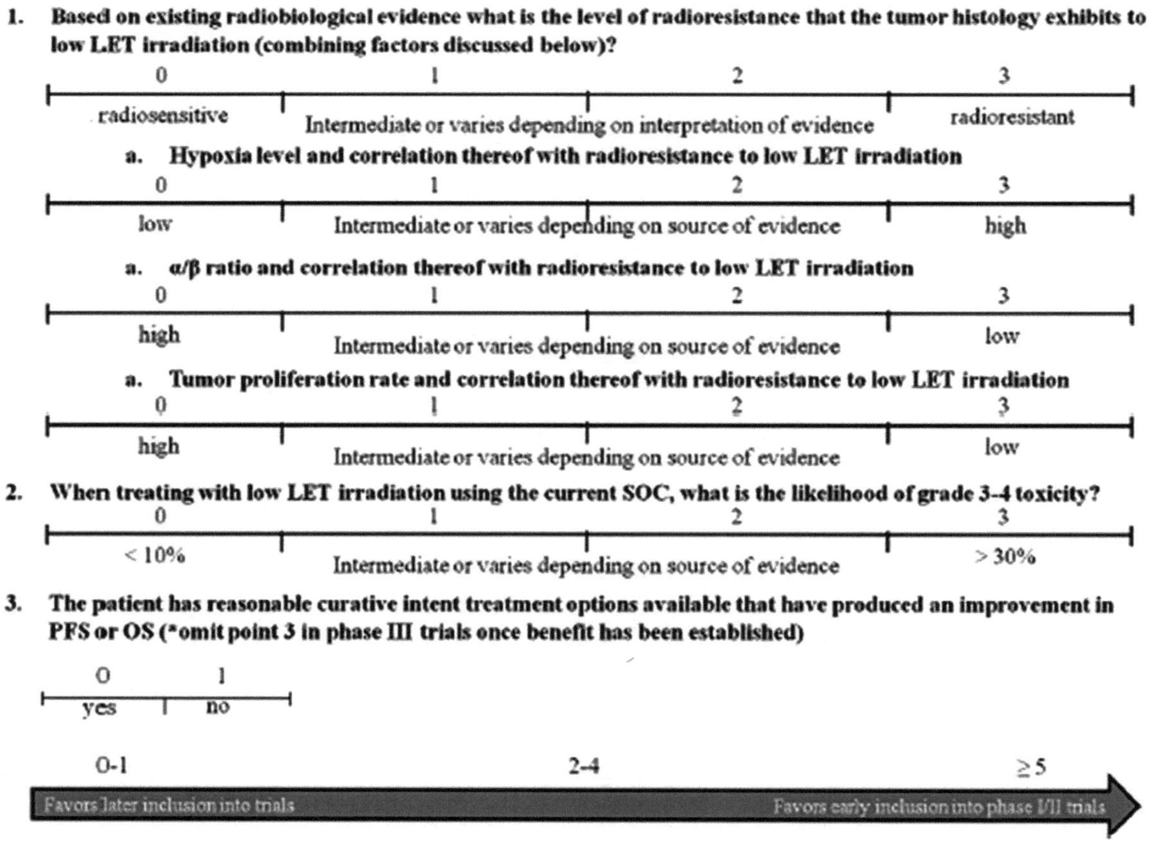

Fig. 5.1 Grading scale of histologies to warrant carbon ion exploration in future clinical trials. Grading scale that should be used to select patients and tumor histologies to determine inclusion earlier or later in clinical trials. *Courtesy* Springer

those tumors that respond well to other radiation species should continue to be treated with these species, while a rare malignancy or patients who do not respond should be treated with carbon ions in an individualized way.

5.5 Clinical Results

Carbon ion therapy has demonstrated benefits for the following:

1. Adenocarcinoma
2. Adenoid cystic carcinoma (ACC)
3. Malignant melanoma
4. Sarcomas arising in the head-neck and many other sites
5. Chordomas of the skull base and sacrum: Significant improvements have been achieved with proton and carbon ion RT; in long-term observation (10 years), the difference in local control rates became larger for carbon ion therapy.
6. Bone and soft tissue sarcomas, including osteosarcoma, chordoma and many other types of sarcomas rising from head and neck, pelvis, vertebra/paravertebral and retroperitoneal region: These tumors are difficult to treat with surgery and are generally photon-resistant.
7. Postsurgical pelvic recurrence of rectal cancer: Treatment has shown comparable or even better results than those achieved with surgery.
8. Malignant melanoma and cancer of pancreas: A combination of carbon ion radiotherapy and chemotherapy has prevented or delays the development of distant metastases with improved survival and local control.

With the unique properties of carbon ions, treatment can be completed in a shorter period of time and with smaller fractions. Future directions of carbon ion therapy depend on the interaction between radiobiology, radiation oncology, and physics accelerators, which combined with the clinical results make it a very promising technique.

5.6 Clinical Advantages of Carbon Ions

The clinical advantages of carbon ions include improved therapeutic gain, hypofractionated radiotherapy, and potential suppression of metastases, as discussed in the following sections.

5.6.1 Improved Therapeutic Gain

The RBE of high-LET carbon ions is greater than low-LET photons and protons. Radiobiological advantages are expected when using carbon ions, such as decreased radiation damage, suppressed tissue population, reduced OER, and reduced dependence on radiosensitivity cell cycle. These advantages are maximized at the peak region, which, combined with improved physical dose location, may have an important role in improving the therapeutic ratio of carbon ion beams and beams of protons and photons. The RBE of carbon ions is similar to that of fast neutron beams. If we apply to the beams of carbon ions findings to the fast neutron therapy [5], carbon ion therapy appears to be effective against photon-resistant tumors and those located near critical structures [6].

5.6.2 *Hypofractionated Radiotherapy (Without Enhancing Toxicity)*

Because of the physical and biological characteristics of carbon ion therapy, it is possible to perform a hypofractionated radiation in relation to the standard used for photons. Experiments with fast neutron beams have shown that increasing the dose per fraction tends to lower RBE for both tumors and normal tissues [7]; however, RBE for tumors does not decrease as rapidly as RBE for normal tissues. Thus, the therapeutic ratio would increase rather than decrease, even if the dose fraction was increased. Experiments conducted with carbon ions, [8, 9] show similar results; therefore, a hypofractionated scheme may be used in carbon ion therapy without increased toxicity.

5.6.3 *Potential Suppression of Metastases*

As is known, the irradiation of carbon ions induces DNA damage, resulting in a high breakage of double-strand DNA, possibly suppressing the metastasis ability of cancer cells in relation to irradiation by X-rays. This is an advantage of treatment using carbon ions. Further studies are needed to confirm these discoveries. Secondary cancer induction after carbon ion therapy, also remains to be studied because there are not valid clinical data.

5.7 The Risk of Secondary Malignancies

More than 38 centers worldwide have treated more than 100,000 cancer patients with particle therapy. Most patients have been treated using protontherapy, but the use of carbon ions is increasing. Unfortunately, despite recent advances in radiotherapy, there is still the risk of cancer arising in a location that was previously free of disease, caused by the treatment itself and not metastasis. There is always the possibility of developing cancer under these conditions because of secondary neutrons, which are inevitably produced in treatments involving particle beams [10]. Quantifying these risks requires a detailed knowledge of a range of parameters and a multidisciplinary team.

Traditional radiotherapy has been improved as a consequence of the development of IMRT, which has enabled improved targeting of conventional X-rays and a reduction in the radiation dose exposure of healthy normal tissue. It is the state of the art in photonics therapy. However, this technique is less effective than carbon ion therapy (as used at HIT). IMRT requires two to three times more monitoring units to deliver a specific radiation dose to the tumor target, when compared with conformal radiotherapy delivered in three dimensions (3D-CRT). Using IMRT instead of 3D-CRT increases the risk of developing secondary cancer by a factor of approximately 2. It is important to remember that particle therapy beams deposit most of their energy near the end of their tracks in the region of the Bragg peak. This peak is spread out to cover the entire tumor volume, and the dose beyond the tumor is lower than in photon therapy.

The neutrons produced during radiation therapy collide with protons in water and generate additional charged particles that can ionize the surrounding molecules. However, this problem can be addressed by using magnetically scanned beams rather than passively scattered beams. It is worth noting that the characteristics of cancer vary from organ to organ, and there is no evidence that the tumor dose–response curves are the same for different organs.

Ionizing radiation is recognized by the World Health Organization (WHO) as a carcinogen. In regions exposed to high doses, ionizing radiation directly kills cells in the field; however, the resulting tissue inflammation and DNA damage to cells in the normal tissue surrounding the tumor can promote cancer. In children, cancer is fortunately relatively rare; when it occurs, radiotherapy is used to treat children with lymphoma, leukemia, brain tumors, sarcomas, Wilm's tumor, neuroblastoma, and liver cancer [11]. In the radiotherapy of pediatric patients, the primary concern is low-dose exposure to distal organs; for adults, high radiation doses induce inflammation.

To estimate the risk of developing cancer from protons, we must rely entirely on animal and in vitro cell experiments. Estimates of the RBE of neutrons are largely based on animal studies, although atomic bomb survivors have also been exposed to neutrons and some data are available. However, considerable uncertainty remains in predicting the late effects of heavy ions in humans. These ions are effective in inducing inflammation. In general, only the organs in the beam path are exposed to heavy ions, while the distal organs receive scattered neutrons and protons.

There is good epidemiological evidence that radiation therapy can contribute to the long-term survival of children with cancer, but it also causes a high incidence of secondary malignancy among survivors. However, the data suggest that hadron therapy leads to a reduced risk of secondary malignancy as compared with conventional radiotherapy modalities that employ X-rays. When using heavy ions, the radiation dose to healthy normal tissues is very low. In addition, the production of neutrons by these ions is lower than is the case for protons, because fewer ions than protons are needed to achieve the same dose in the tumor target.

The lack of resources on hadron therapy has led to inadequate recommendations of continued conventional radiotherapy in cases of cancer recurrence. This has prevented proper treatment with hadron therapy techniques using carbon ions, leading patients to undergo chemotherapy as the only treatment, which is not always satisfactory.

5.8 Clinical Trials at HIT

Fifteen trials were initiated at HIT since November 14, 2009, when the first patient was treated. A brief description of each trial follows [12]:

1. *HIT-1 trial for chordomas of the skull base* The first trial at HIT was a prospective randomized phase II trial for the treatment of skull base chordomas. This trial tested for superiority of carbon ion irradiation against proton irradiation with respect to the local progression-free survival. The 5-year local progression-free survival was 70 % using protons and 80 % using carbon ions.
2. *HIT chondrosarcoma trial* This trial is also a prospective randomized phase II trial for the optimal treatment of skull base chondrosarcomas.
3. *COSMIC trial* The goal of the trial was to evaluate toxicity in dose-escalated treatment with intensity-modulated radiotherapy (IMRT) and carbon ion boost for malignant salivary gland tumors of the head and neck.
4. *ACCEPT trial* The ACCEPT trial followed the COSMIC trial in the treatment of adenoid cystic carcinomas (ACC) of the head and neck. However, in this trial, only ACC patients with microscopic residual disease can be included.
5. *IMRT-HIT-SNT trial* This trial examines the effect of a carbon ion boost in the treatment of patients with unresected or incompletely resected nasal or paranasal sinus carcinomas.
6. *TPF C-HIT trial* Locally advanced tumors of the oropharynx, hypopharynx, and larynx are suitable for this trial. Except for the induction chemotherapy using docetaxel, cisplatinum and 5FU (TPF), the design is similar to the ACCEPT trial.
7. *CLEOPATRA trial* This trial is a single-center randomized phase II trial for the treatment of glioblastomas. The aim of the trial is to show the overall survival (primary end point) for glioblastoma patients using a carbon ion boost dose escalation compared to the standard treatment.
8. *CINDERELLA trial* The effect of carbon ion irradiation in the treatment of recurrent gliomas after initial radiation treatment is examined in the CINDERELLA trial.
9. *MARCIE trial* Atypical meningiomas have a much higher recurrence rate than meningiomas of WHO grade I. The Phase II MARCIE study evaluates a carbon ion boost applied to the macroscopic tumor in conjunction with photon

radiotherapy in patients with atypical meningiomas after incomplete resection or biopsy.

10. *IPI trial* The role of the use of ions in the primary treatment of prostate cancer is unknown. There are no prospective proton data. However, NIRS has published promising results in respect of the hypofractionated use of carbon ions. The IPI trial wants to confirm these Japanese data in a prospective randomized phase II trial.
11. *PROLOG trial* This trial focuses on the use of protons in the postoperative situation of prostate cancer (either as adjuvant treatment or as salvage treatment).
12. *PANDORA trial* Patients with recurrent rectal cancer are still challenging. This trial examines the role of carbon ions in the reirradiation of patients with recurrent rectal cancer.
13. *ISAC trial* Imai et al. published the NIRS data with respect to hypofrationated carbon ion irradiation of sacral chordoma. This trial will confirm these data using the raster scan method. Additionally, it will be examined if these results are an effect of the use of carbon ions or an effect of high and hypofractionation.
14. *OSCAR trial* This trial is a non-randomized therapy trial to determine the safety and efficacy of heavy ion therapy in patients with nonresectable osteosarcomas. The primary endpoint of OSCAR is feasibility and toxicity in the ion treatment of unresectable osteosarcoma.
15. *PROMETHEUS trial* The PROMETHEUS trial is the first trial evaluating carbon ion radiotherapy delivered by intensity-modulated raster scanning for the treatment of hepatocellular carcinoma.

Assays were initiated to assess the role of carbon ions and protons in the treatment of a variety of cancer types. In the future, these will be included in HIT trials with moving targets. For example, the INKA test will start neoadjuvant radiotherapy using raster scanned carbon ions in patients with locally advanced sulcus superior tumors. The role of ions for radiotherapy in the treatment of pancreatic cancer will be developed, as well as the use of carbon ions for inoperable esophageal cancer [12].

5.9 Consolidated, Prospective, and Exceptional Indications Using Carbon Ion Therapy

5.9.1 Consolidated Indications

Consolidated indications (Table 5.2) are the core indications that have been treated effectively using neutron therapy (salivary gland tumors, adenoid cystic carcinomas of the upper respiratory and digestive tracts, particularly the trachea, superficial sarcomas) and are currently treated in Japan and Germany (adenocarcinomas of the head and neck, mucosal melanomas, chordomas, sarcomas, hepatocellular

Table 5.2 Consolidated indications resulting from Etoile's work

Tumor location	Detailed definition of indications	Recommended form of hadron therapy	Estimated incidence[a] (cases/year in France)
Salivary gland (parotid gland) tumours	Inoperable tumours or refusal of surgery or R2 restrictions or local recurrences[b]	Carbon alone or in combination with a close of locoregional photon therapy	=100
	All types of histology: adenoid cystic carcinomas, mucoepidermoid adenocarcinomas, acinar cell carcinomas, etc.		
Paranasal sinus tumours	Inoperable tumours or refusal of surgery or R2 restrictions or local recurrences	Carbon alone in primary location	=250
	Adenocarcinomas and adenoid cystic carcinomas		
Adenoid cystic carcinomas with skull base involvement	Inoperable tumours or refusal of surgery or R2 resections or local recurrences	Carbon alone in primary location	=10
Malignant mucosal melanomas (primarily ENT)	Any location without immediately threatening metastasis	Carbon alone in primary location; urgent treatment	=40
	Tumour without surgery if possible or emergency after R2 resections or non-irradiated local recurrence		
Chordomas at the base of the skull, spine and sacrum	Any clinical presentation	Carbon or protontherapy alone in primary location	=30–50
Chondrosarcomas of the axial skeletal	Base of skull	Protontherapy alone in primary location	=20
	Spine and sacrum	Protontherapy or carbon alone in primary location	<10
			=100

(continued)

Table 5.2 (continued)

Tumor location	Detailed definition of indications	Recommended form of hadron therapy	Estimated incidence[a] (cases/year in France)
Soft-tissue (non-retroperitoneal) sarcomas	Weak-grade M0, any histology, any location	Carbon alone in primary location	
	Unresectable or surgery refused or "definitive R2": R2 with no possible repeat surgery or R2 following repeat surgery or local recurrence in R2 resection		
	Non-threatening M+ with incapacitating T or rT		=80
Retroperitoneal sarcomas	Following local recurrence and surgical resection: R0 or R1 and M0 (for unresectable T and R2, see above)		=40
	Initial status R1 M0		
Soft-tissue sarcomas of the head, neck and limbs	"Definitive R1": R1 resection with no acceptable possibility for repeat surgery		=200
Osteo- and chondrosacromas (any location except axial skeleton)	Tumours without surgery or resection: R2, M0		=10
	M+ accepted for osteosarcomas only		
	Discussion according to grade		
Pelvic recurrence of rectal adenocarcinomas	Unresectable unifocal locoregional pelvic recurrence in irradiated or	Carbon alone	=200
	non-irradiated location, and M0 (CT, liver MRI, PET)		
Hepatocellular carcinomas	Single hepatocellular carcinoma, a > 4–5 cm, unresectable, M0, not suitable for conventional treatment methods or photon therapy, no threatening comorbidity	Carbon alone in primary location	=50

Courtesy Centre Etoile, Groupement de Coopėration Sanitaire. Centre National D'Hadrontherapie par ions Carbone [13]

Notes

[a]The annual estimated incidence is the estimated total annual number of tumours that match the detailed descriptions. This is the maximum recruitment potential. It does not take into account feasibility of treatment or the care services actually available

[b]Local recurrence is taken to mean the reappearance of the tumour in its primary location, with no other regional or metastatic manifestation

carcinomas, pelvic recurrences of rectal adenocarcinomas). Their published outcomes are well above the figures obtained using non-carbon ion therapy (approximately 20–25 % higher for 5-year local control) [13].

5.9.2 Prospective Indications

Prospective indications are shown in Table 5.3.

Table 5.3 Prospective indications resulting from Etoile's work

Tumour location	Detailed definition of indications	Recommended form of hadron therapy	Estimated incidence[a] (cases/year in France)
Nan-small cell lung cancer	Inoperable initial, status (UICC/AJC 1997) IA and 1B: T1T2N0 (CT, PET) M0 (brain MRI): purely endobronchial tumours excluded	Carbon alone in primary location with respiratory gating	=750–1000
	Second cancer in patients who underwent radiotherapy and/or pneumonectomy >2 years ago; inoperable stage I		
	Inoperable initial status, stage (UICC/AJC 1997) IIB-IIIB limited to T3T4No (CT, PET) M0 (brain MRI); purely endobronchial tumours excluded		
	Second cancer in patients who underwent radiotherapy and/or pneumonectomy >2 years ago; inoperable stage II		
Nasopharynx	Any histology	Protons or carbon	=10
	Strictly local recurrences[b] after initial radiation		

(continued)

Table 5.3 (continued)

Tumour location	Detailed definition of indications	Recommended form of hadron therapy	Estimated incidence[a] (cases/year in France)
High-grade gliomas (grade 3 or glioblastomas)	Recurrence after initial radiotherapy ± chemotherapy and progressing during chemotherapy	Carbon alone in primary location	=50
	Initial treatment, possibly following surgery		=300
Epidermoid ENT carcinomas	Unresectable recurrences or second location, in irradiated area and M0(CT,liver MRI, PET)(*proposal to be assessed*)	Carbon alone	=500
	Initial status T3–T4, N ≤ 2, M0 of the oropharynx or oral cavity (*proposal to be assessed*)		
Prostate adenocarcinomas	Intermediate risk groups:	Comparison IMRT ± hormone therapy versus carbon versus protons	=1000
	T2b,T3a/b and (PSA 10–20 and/or Gleason ≥ 7) and PN0)		
Highly radioresistant tumours of digestive tract	Unresectable single nodular bile duct cancer or pancreatic adenocarcinoma, M0, not previously irradiated and not progressing during chemotherapy after 4–6 months	Carbon alone or in combination with dose of locoregional photon therapy	=900
	M0 endocrine tumour of the pancreas, progressing after multiple treatments: Isotopic and/or chemotherapy and somatostain	Carbon alone in primary location	=20

Courtesy Centre Etoile, Groupement de Coopération Sanitaire. Centre National D'Hadrontherapie par ions carbone [13]

Notes

[a]The annual estimated incidence is the estimated total number of tumours that match the detailed descriptions. This is the maximum recruitment potential. It does no take into account feasibility of treatment or the care services actually available

[b]Local recurrence is taken to mean the reappearance of the turnour in its primary location, with no other regional or metastatic manifestation

5.9.3 Exceptional Indications

Exceptional indications (Table 5.4) are absolutely isolated situations with no other treatment options [13]. They include both pediatric indications that have not yet been treated using carbon ion therapy but are clearly life-threatening, as well as exceptionally rare, radioresistant tumors that are strictly threatening to locoregional or vital functions. These indications can certainly never be the subject of comparative studies, but nevertheless they must be discussed by experts at multidisciplinary consultation meetings and be traceable in terms of management by carbon ion therapy if this is authorized.

Table 5.4 Exceptional indications proposed by Etoile

Tumor location	Detailed definition of indications	Recommended form of hadron therapy	Estimated incidence (cases/year in France)
Paediatric tumours	Large (more than 100 or 200 ml, depending on age), inoperable	Carbon alone in primary location	<100
	Ewing's sarcomas of the pelvis		
	Aggressive chordomas in small children(<3–4 years)		
	Unresectable pelvic osteosarcomas		
Various locations, highly functional	Benign tumours or locally-invasive malignant tumours that are incapacitating and have a high risk of local recurrence (desmoid tumours, neurinomas, schwannomas, meningiomas, etc.)	Carbon alone	Very rare

Courtesy Centre Etoile, Groupement de Coopération Sanitaire, Centre National D'Hadrontherapie par ions carbone [13]

The ETOILE Centre is managed according to the underlying principle of achieving financial viability by treating "consolidated" indications alone. As a result, this document will examine only these indications in terms of the comparative results of carbon ion therapy

However, publications on other indications (prostate cancers, lung cancers, gliomas, pancreatic cancers) are growing in number. This may alter some priorities over time

5.10 New Cancers Where Charged Particles May Potentially Lead to a Breakthrough

Figure 5.2 shows an estimate of new cancer cases and estimated cancer deaths in the United States in 2013 [14]. New cancers where charged particles may potentially lead to a breakthrough include the following [15]:

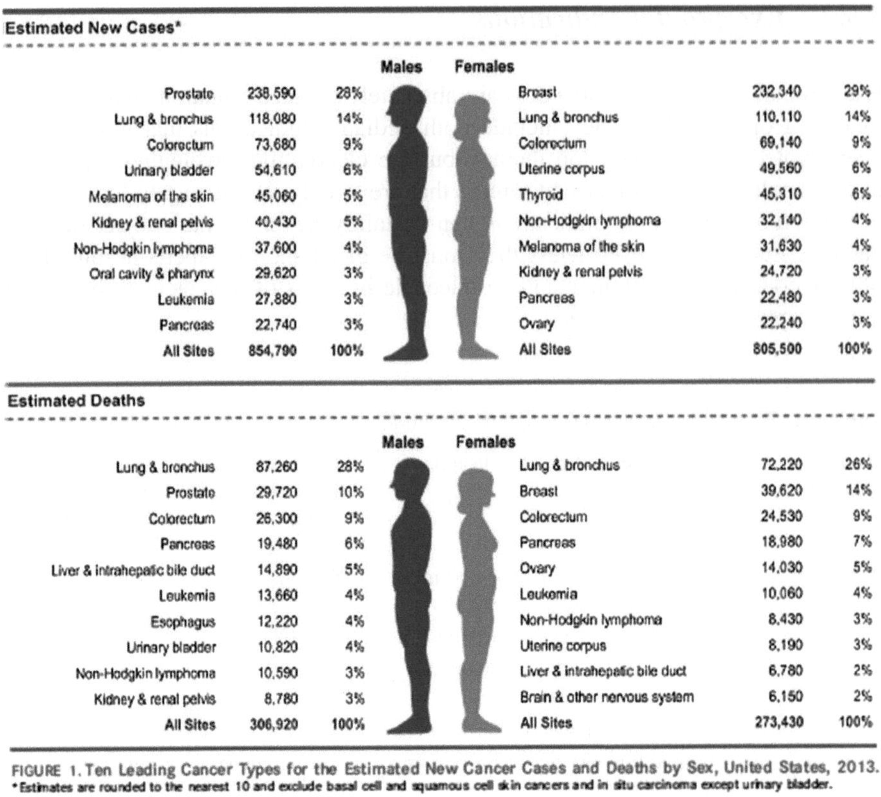

Estimated New Cases*

Males			Females		
Prostate	238,590	28%	Breast	232,340	29%
Lung & bronchus	118,080	14%	Lung & bronchus	110,110	14%
Colorectum	73,680	9%	Colorectum	69,140	9%
Urinary bladder	54,610	6%	Uterine corpus	49,560	6%
Melanoma of the skin	45,060	5%	Thyroid	45,310	6%
Kidney & renal pelvis	40,430	5%	Non-Hodgkin lymphoma	32,140	4%
Non-Hodgkin lymphoma	37,600	4%	Melanoma of the skin	31,630	4%
Oral cavity & pharynx	29,620	3%	Kidney & renal pelvis	24,720	3%
Leukemia	27,880	3%	Pancreas	22,480	3%
Pancreas	22,740	3%	Ovary	22,240	3%
All Sites	854,790	100%	All Sites	805,500	100%

Estimated Deaths

Males			Females		
Lung & bronchus	87,260	28%	Lung & bronchus	72,220	26%
Prostate	29,720	10%	Breast	39,620	14%
Colorectum	26,300	9%	Colorectum	24,530	9%
Pancreas	19,480	6%	Pancreas	18,980	7%
Liver & intrahepatic bile duct	14,890	5%	Ovary	14,030	5%
Leukemia	13,660	4%	Leukemia	10,060	4%
Esophagus	12,220	4%	Non-Hodgkin lymphoma	8,430	3%
Urinary bladder	10,820	4%	Uterine corpus	8,190	3%
Non-Hodgkin lymphoma	10,590	3%	Liver & intrahepatic bile duct	6,780	2%
Kidney & renal pelvis	8,780	3%	Brain & other nervous system	6,150	2%
All Sites	306,920	100%	All Sites	273,430	100%

FIGURE 1. Ten Leading Cancer Types for the Estimated New Cancer Cases and Deaths by Sex, United States, 2013.
*Estimates are rounded to the nearest 10 and exclude basal cell and squamous cell skin cancers and in situ carcinoma except urinary bladder.

Fig. 5.2 Leading cancer types for estimated new cancer cases and deaths by sex, United States, 2013. From Siegel et al., CA Cancer J Clin 2013. *Courtesy* John Wiley and Sons

- Lung
- Pancreas
- Local recurrence of rectal cancer
- Breast
- Hepatocellular carcinoma
- Glioblastoma

Certainly, it is recommended to combine treatments to improve survival rates.

5.11 Protontherapy Versus Carbon Ion Therapy: Advantages, Disadvantages, and Similarities

Comparative studies between protontherapy and carbon ion therapy are limited for several reasons:

1. Multiple institutions
2. Dose fractionation differences, making it impossible to compare the efficacy of protontherapy versus carbon ion therapy
3. Different populations of patients for the disease in question (e.g., patient gender, age, health condition)
4. A small number of patients in each modality
5. Absence of an experimental protocol to be used in the various institutions
6. Lack of establishment of a variable, such as the LET, for the study of comparative effects between the techniques

However, both protontherapy and carbon ion therapy are particulate forms of therapy that can be effectively used to treat tumors. Studies show an approximate equivalence of the two therapies. A more detailed analysis is provided here.

The **advantages of protontherapy over carbon ion therapy** include the following:

1. Protontherapy requires less expensive equipment and facilities than carbon ion therapy—approximately one half or one third of the price of a system for carbon ion therapy. Currently, costs are dropping, with more compact gantries. The commercial supply of hybrid systems allows results using both techniques.
2. Protontherapy allows the use of the gantry, providing its beam's positioning in multiple angles. Carbon ion therapy usually works with fixed angles, not allowing multiple angles. However, this problem was solved at HIT, with a 670-ton gantry; for the C400 IBA at JINR, studies are in progress, and IBA has already developed a compact gantry (ProteusOne) with success [16]. The use of a gantry is essential for both therapies.
3. Protontherapy has a narrower RBE (in the range of 1–1.1) and therefore greater certainty, leading to minor variations of the released current dose. Carbon ion therapy's RBE is 1.5–3.4, which can cause large variations in the current dose released [17].
4. Due to its low RBE, protontherapy decreases the risk of a late injury to normal tissue. Carbon ion therapy, due to its high-value and widely varying RBE, has the potential to increase the risk of late damage to normal tissue [18].

The **advantages of carbon ion therapy over protontherapy** include the following:

1. Treatment by carbon ions deposits 24 times more energy in tumor cells than protontherapy [19].
2. The use of carbon ions is about 52 times more accurate than the use of protons with regard to the precise control of the depth of penetration of the particle [20]. This is an important because conventional radiotherapy—even using the IMRT, which exhibits excellent conformity—does not exceed the accuracy achieved by the technique using particles.
3. Carbon ion therapy is ideal for the treatment of deep-seated tumors, where the penumbra becomes a limiting factor. This can be explained by the greater mass of carbon ions, producing three times less multiple scattering and range

straggling compared to protons, causing a sharp side edge and longitudinal. The penumbra [80–20 %] is the dose width sideband for the field edge to decrease the dose from 80 to 20 %. [80–20 %] for a collimated beam passively scattered proton varies markedly with depth, being highly dependent on the physical design of the systems that define the beam machine, collimation system, and air space (air gap) between the compensator and the body's surface [21].

4. Carbon ions have increased therapeutic benefits in treating tumors resistant to radiation. The linear energy transfer at the peak of a carbon beam is higher than that for beams of photons and protons, with a relative biological effectiveness (RBE) that is two to three times higher for carbon ions. Carbon ion therapy is three times more efficient than protontherapy [22].
5. Compared to protontherapy, carbon ion therapy shows a detectable level of activation with less uncertainty, turning the online positron emission tomography easier and having the advantage of three-dimensional treatment verification. Observing the Bragg peak produced by carbon ions, there is the presence of a tail in the distribution of the dose due to fragmentation of carbon ions, contributing to a distal small dose (Fig. 5.3). These carbon fragments are indicative of the production of a large number of local positron emitters, in global volume under treatment [6].

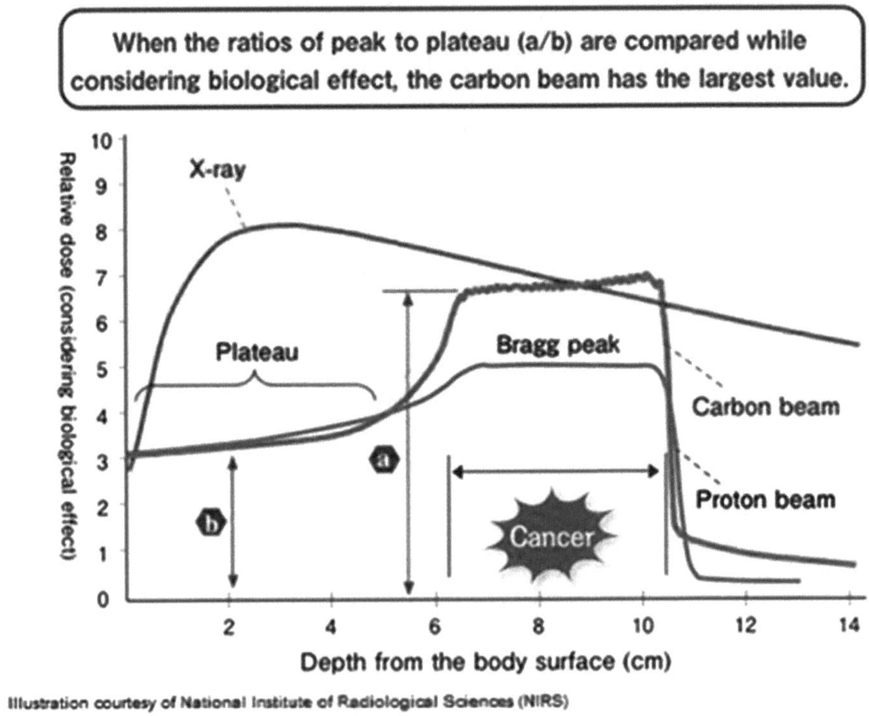

Fig. 5.3 Dose distribution of radiation considering biological effects. *Courtesy* National Institute of Radiological Sciences (NIRS)

6. Because of its higher RBE value, especially at the distal end of the Bragg peak, carbon ion therapy allows greater tumor control. Protontherapy has a similar RBE to conventional radiation photons and therefore it is not expected to increase tumor control.
7. Carbon ion therapy has a smaller lateral penumbra, allowing a conformational dose (laterally) and limiting the damage to normal tissue. Protontherapy has a large lateral penumbra, which can cause a higher dose to normal tissue structures in comparison to carbon ion therapy. Therefore, the dose distribution of a single beam appears to be better in carbon ion therapy than in protontherapy [1].
8. In cases of cancer relapse or a pediatric tumor, the indicated therapy is by carbon ions [19].
9. Carbon ions have some clear advantages over protons in providing both local control of very aggressive tumors and lower acute or late toxicity, thus enhancing the quality of life during and after cancer treatment [13].
10. Based on Tables 5.2, 5.3 and 5.4 for the consolidated, prospective, and exceptional indications offered by the Etoile Group, the recommended hadron therapy has always been carbon ion therapy.

Similarities Between the Two Techniques

1. Both protontherapy and carbon ion therapy reduce the risk of secondary malignancies compared to conventional photon therapy because they limit the full dose. This is particularly important in pediatric populations.
2. Both therapies are limited in research, consisting of a small series of patients, so definitive conclusions are difficult to make.

Conclusion

Carbon ions have a higher LET (75–300 keV/μm) than photons, electrons, or protons (0.2–5), resulting in a dense ionization along its track and causing more irreparable harm, both by direct and indirect means. In the indirect route, there is free radical ionization of water. Thus, the damage to DNA helices is greater. Less oxygen dependency and higher potential for hypoxic or anoxic targets occur (i.e., radio-resistant tumors). Carbon ions also have cell-cycle independent kill, as opposed to low-LET radiation.

In summary, compared with photons and protons, carbon ions exhibit the biophysical advantages of narrow penumbra, higher LET, higher RBE (1–1.1 vs. 2–3), less dependence on oxygen, less dependence on cell cycle, and therapy requiring hypofractionation. Carbon ions appear to be a promising modality in oncology radiation, and there has certainly been an international effort to establish their effectiveness using well-conducted experiments. In the near future, the full potential of ions and their clinical applications will be better defined.

5.12 What Do We Need?

The following are needed for the future of hadron therapy:

1. The low energy ionizing ring (Fig. 5.4) is needed. LEIR, which is CERN's biomedical facility for basic physical studies, radiobiology, fragmentation of the carbon ion, dosimetry, and test instrumentation, was requested by ENLIGHT (a community with more than 20 participating countries, involving more than 200 people). It is a feature of Europe for radiobiology [23].
2. An expansion in the clinical indications for the spread and effective use of hadron therapy is needed. The lack of medical recommendations for the use of hadron therapy often results from a lack of knowledge about the excellence and precision of technique, not the conservatism of doctors.
3. Medical physics should reduce the uncertainty of range and treatment time, making it possible to treat moving parts (gating), oligometastasis, etc.
4. More elaborate studies about hypofractioning are needed.
5. Combined treatments with techniques in development, such as vectorial chemotherapy and nanotherapy, will be useful.
6. For highly lethal and incurable diseases, radiobiology can lead to breakthroughs in cancer therapy.
7. The physics of accelerators should reduce costs and include compact gantries, compact accelerators, and innovative approaches such as as the cyclinac, DWA, and laser.
8. Protontherapy and carbon ion therapy should be optimized as follows:
   (a) For protontherapy, the scanning pencil beam and intensity modulated protontherapy (IMPT) should be improved.
   (b) For carbon ion therapy, inquiries regarding RBE and the isoeffective dose should be made.
9. Research is required to select protontherapy versus carbon ion therapy based on the alpha/beta ratio values of the target and structures of the normal tissue around it.
10. Results of randomized trials developed at HIT—a total of 15 related to carbon ion therapy—are needed.
11. A great percentage of radiation treatment will use particle beams and four-dimensional image-guided radiation therapy with tracking and beam repositioning to maintain the target correctly positioned to the beam.
12. The high LET of carbon ion beams is a clinical advantage, but the magnitude of this advantage is unknown.
13. Studies are required on the clinical gain of a narrower penumbra.

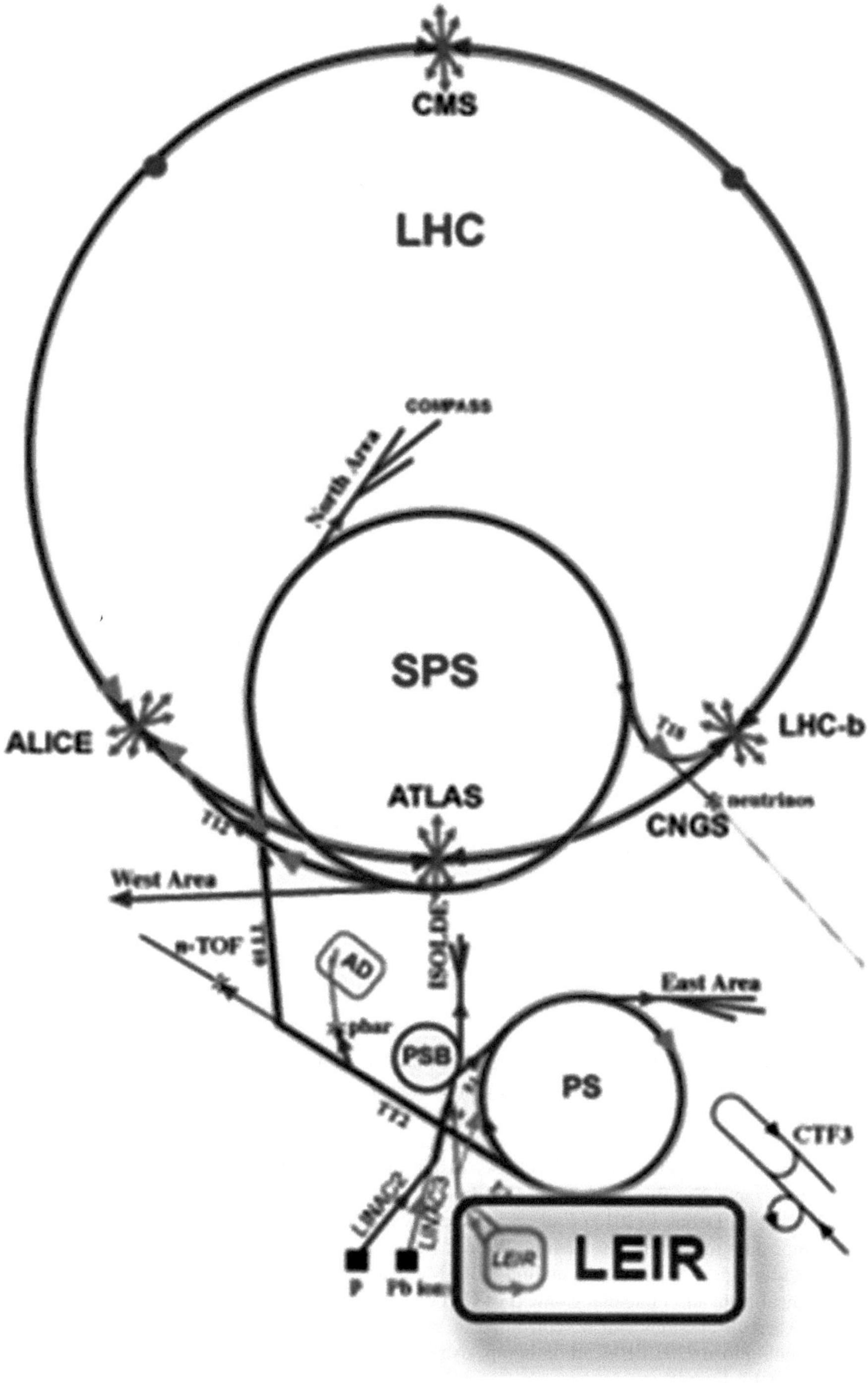

Fig. 5.4 The low energy ionizing ring (LEIR) at CERN. *Courtesy* CERN

Regarding the creation of an organizational structure, promotion and media

1. An organizational framework for international coordination of hadron therapy should be created, with the establishment of research standards, new protocols, etc. This will allow the comparison of clinical outcomes of randomized trials with different techniques, such as protontherapy and carbon ion therapy.
2. Financial support should be expanded for international collaboration in opening new centers of research and clinical treatment in hadron therapy.
3. Research scholarships should be increased by the international funding agencies for researchers who are interested in working with hadron therapy.
4. An enhanced international exchange of scientists is needed, generating courses, conferences, symposiums, and media coverage.

The current approaches to treatment strategies will be considered obsolete in the future. It is our responsibility to contribute so that such a future can be reached or exceeded.

References

1. Chen GTY, Castro JR, Quivey JM (1981) Heavy charged particle radiotherapy. Ann Rev Biophys Bioeng 10:499–529
2. Ohara K, Okumura T, Akisada M et al (1989) Irradiation synchronized with the respiration gate. Int J Radiat Oncol 17:853–857
3. Amaldi U, Kraft G (2007) European developments in radiotherapy with beams of large radiobiological effectiveness. J Radiat Res (Tokyo) 48(Suppl A):A27–A41
4. Schlaff CD, Krause A, Belard A et al (2014) Bringing the heavy: carbon ion therapy in the radiobiological and clinical context. Radiat Oncol 9:88–107
5. Lamore GE (1997) The use of neutrons in cancer therapy: a historical perspective through the modern era. Semin Oncol 24:672–685
6. Tsujii H, Kamada T (2012) A review of update clinical results of carbon ion radiotherapy. Jpn J Clin Oncol 42(8):670–685
7. Denekamp J, Waites T, Fowler JF (1997) Predicting realistic RBE values for clinically relevant radiotherapy schedules. Int J Radiat Biol 71:681–694
8. Ando K, Kase Y (2009) Biological characteristics of carbon-ion therapy. Int J Radiat Biol 85:715–728
9. Ando K, Koike S, Usawa A et al (2005) Biological gain of carbon-ion radiotherapy for the early response of tumor growth delay and against early response of skin reaction in mice. J Radiat Res 46:51–57
10. Timlin C, Jones B (2010) Proton and charged particle radiotherapy. Br J Radiol 83:87
11. Newhauser WD, Durante M (2011) Assessing the risk of second malignancies after modern radiotherapy. Nat Rev Cancer 11:438–448
12. Herfarth K, Debus J (2013) Carbon ion therapy: actual and future strategies at HIT. In: Joint Symposium on carbon ion therapy. Wiener Neustadt, Austria. Organized by National Institute of Radiological Sciences, Japan and EBG MedAustron GMbH, Austria
13. Medical Data and Financial Context of Carbon Ion Therapy (2013) Study files and summaries of expert discussions can be consulted at www.centre-etoile.org. http://centre-etoile.org/wp-content/uploads/2013/10/Carbontherapy-An-update-of-medical-and-economic-data-as-for-June-2013_En.pdf

14. Siegel R, Naishadham D, Jemal A (2013) Cancer statistics-2013 CA: a Cancer. J Clinicians 63:11–30. http://onlinelibrary.wiley.com/doi/10.3322/caac.21166/full
15. Durante M (2014) Radiobiology for particle therapy. CNAO-NIRS meeting, Pavia 21.03.2010. INFN Workshop. Napoli, 4.4.2014
16. http://www.medgadget.com/2014/07/new-compact-gantry-for-proteusone-proton-therapy-system-fda-cleared-video.html
17. Mihailescu D, Borcia C (2012) Biophysical models in hadrontherapy. J Adv Res Phys 3(1):1–9
18. Tsujii H, Kamada T, Shrai T et al (2013) Carbon-Ion radiotherapy: principles, practices, and treatment planning. Springer Japan
19. Nunes MA (2013) Hadron therapy physics and simulations. Springer, New York
20. Trbojevic D et al (2011). Lattice design of a rapid cycling medical synchrotron for carbon/proton therapy. In: Proceedings of IPAC 2011, San Sebastian, Spain. http://www.teambest.com/news/BPT_CarbonIonSynchrotron_4page_Oct2014_MedPhys1.pdf
21. Suit H, DeLaney T, Goldberg S et al (2010) Proton vs carbon ion beams in the definitive radiation treatment of cancer patients. Radiother Oncol 95:3–22
22. Amaldi U, Bonomi R, Braccini S et al (2010) Accelerators for hadrontherapy: from Lawrence cyclotrons to linacs. Nuclear Instruments and Methods in Physics Research A 620:563–577
23. http://medicalphysicsweb.org/cws/article/opinion/56295

Appendix

Questions and Answers About Proton Versus Carbon Ion Therapy

1. Why is hadron therapy not recommended as the new frontier of cancer treatment by doctors around the world?
   Due to the price of the equipment, there is great difficulty in its acquisition and therefore it is more difficult to expand the use of this new technique. In addition, there is a lack of knowledge about hadron therapy among physicians and the general population.
2. Is there a difference between the results obtained for the RBE using the LEM models of the GSI/HIT and the model used at HIMAC?
   Yes. GSI/HIT uses biophysical modelling and a local effect model (LEM), with a variable RBE scheme. HIMAC uses experimental data based on clinical experiments with neutrons and fixed RBE. The difference in results determined by the two models is significant and reaches 15 %. There is therefore a need to introduce a correction factor when comparing the two models.
3. What is currently the most accepted model for determining the RBE?
   The most widely accepted model for determining the RBE is the local effect model (LEM), which was developed by Scholz and Kraft (Department of Biophysics of GSI) and is used at HIT and CNAO.
4. Which models are used at CNAO?
   At CNAO (Pavia, Italy), two models are used: the local effect model (LEM) and the microdosimetric kinetic model (MKM).
5. Does the RBE increase as the LET increases?
   RBE increases with increasing LET along the same SOBP. However, we must specify a LET range for which this is valid (10–120 keV/μm, in various experimental conditions).
6. What does a high alpha/beta ratio mean? A low alpha/beta ratio?
   A high alpha/beta ratio is normally found in human tumors, suggesting the predominance of an alpha component—implying decreasing response to the fractionation and therefore the clinical benefit of hyperfractionation. Hyperfractionation is implemented in order to prevent accelerated repopulation and maximizes therapeutic gain. A low alpha/beta ratio (1.5–5 Cy) is usually

M. d'Ávila Nunes, *Protontherapy Versus Carbon Ion Therapy*,
Biological and Medical Physics, Biomedical Engineering,
DOI 10.1007/978-3-319-18983-3

associated with delayed response of normal tissue and is the basis for therapeutic gain using hypofractionation. Some tumors have a low alpha/beta ratio, including prostate cancer, rhabdomyosarcoma, and melanoma.

7. Where is the birthplace of the use of carbon ions in hadron therapy?
   The initial experiments were developed at Lawrence Berkeley National Laboratory (LBNL) in Berkeley, CA, USA in 1954. Japanese researchers interning at Berkeley carried the technique to Chiba, Japan and since then only worked on the development and clinical applications of carbon ion therapy and protontherapy. Later, the Germans brought the technique to Darmstadt, the GSI, and from there to HIT.
8. What is gating? When it should be used?
   When therapy is carried by protons or carbon ions of moving organs, there is a need for the radiation beam to remain within the boundaries of the tumor; thus, the gating technique must be used. Gating has been used more for treatment of lung cancer and hepatocellular injury.
9. What is passive and active scanning?
   Because the Bragg peak is too sharp and thin to be used directly in treating tumors of different shapes and sizes, peak broadening is necessary to conform to the size and shape of the tumor. A beam scattering method can be used with a passive system of beam release or beam scanning method with an active system of beam release. In the passive system, the narrow peaks are swept over a wide area by a peak filter to create a spread-out Bragg peak (SOBP), corresponding to the target volume size. This method used a combination of a band modulator, collimator, and compensator. In the active system, the peak position is moved within the target due to variation of beam energy in the accelerator or changing the beam's penetration using absorbers, at a dose sufficient to conform precisely to the target volume.
10. How many centers currently exist in the world that process carbon ion therapy? How many are under construction? Are there carbon ion therapy centers in the United States?
    Carbon ion therapy is conducted in five facilities in the world. In addition to HIMAC/NIRS in Japan; GSI in Darmstadt, Germany; and the HIT in Heidelberg, Germany, the Hyogo Ion Beam Medical Center was created in 2001 to focus on protons and carbon ions. Since 2006, the Institute of Modern Physics in Lanzhou, China, has developed clinical trials using carbon ions for the treatment of superficial tumors with a system of only 100 MeV/u. In 2010, Gunma University Heavy Ion Medical Center in Japan has conducted clinical studies. In 2011, the CNAO was created in Pavia, Italy, providing treatment with protons and carbon ions and employing an excellent synchrotron to accelerate carbon ions as of 2012. In Germany, two centers are under construction in Marburg and Kiel, with similar systems as the HIT. A center is also under construction in Shanghai, China. Six institutions are building new centers: two in Japan, one in Austria, one in China, and two in the United States (Colorado and Michigan).

11. How much is needed to build a center for carbon ions? Can the cyclotron be used for carbon ion therapy? What model and specifications are recommended?
    Approximately $200 million is needed to build a new center (this is the amount used by the University of Colorado to build their new carbon ion therapy center). If a cyclotron is a C400 type, it can be used by working with superconductors and processing protontherapy and carbon ion therapy in a single instrument. The cyclotron must offer at least 400 MeV/u, 0.1 nA current.
12. Why does the synchrotron at Loma Linda University Medical Center not process carbon ion therapy, only protontherapy?
    The synchrotron of the Loma Linda University Medical Center has a energy of only 250 MeV/u, allowing only protontherapy.
13. Why are carbon ions used and not bigger ions?
    Lawrence Berkeley National Laboratory tested ions that were larger than carbon ions (neon). However, the carbon ions provide the best balance between the properties of various ion species in terms of physical dose distribution and biological effects.
14. What are the new development techniques to process hadron therapy?
    The new techniques are laser and dielectric wall accelerator (DWA). In the case of laser, a powerful laser pulse that acts on a target formed by a thin lamina doped with hydrogen atoms is applied. The laser accelerates the electrons out of the target in its back, creating a favorable electric field output of protons. The DWA is a class of induction accelerators. Electrodes are placed as sandwiches between insulating plates of high gradients and subjected to a high traveler field. The accelerator tube is made of fused silica of 250 microns thick, which is a pure transparent quartz—an insulator (hence the DWA name). Further details can be found in the text.
15. What are the similarities between the two techniques using protons and carbon ions?
    Both protontherapy and carbon therapy reduce the risk of secondary malignancy compared to conventional photon therapy because of their integral dose limit. This is important in the pediatric population. Studies of both protontherapy and carbon ion therapy are limited, consisting of small series of patients; thus, definitive conclusions are difficult to make.
16. When we should use carbon ion therapy instead of protontherapy or IMRT photon?
    The use of carbon ion therapy is recommended when the advantages of using carbon ions outweigh the therapeutic advantages that can already be obtained with photon RT fractionation.
17. How many clinical trials are currently underway at HIT?
    A total of 15 randomized clinical trials are currently underway at HIT. See text for details of these trials.

18. What is the range of alpha/beta ratios by biological system (tumors)? What does this mean?
    The range is 0.4–16, with 0.4 for the nasopharynx and 16 for liposarcoma. A high alpha/beta ratio high (6–14 Gy), seen in many human tumors, suggests a predominance of the alpha component, implying a decreased response to fractionation and thus the clinical benefit of hyperfractionation. A low alpha/beta ratio (1.5–5 Gy) is usually associated with delayed response of normal tissue and is the basis for the therapeutic gain achieved by using hypofractionation.
19. Is there a center for hadron therapy in South America or Central America?
    There are no protontherapy centers or carbon ion therapy center in South America. In Central America, a protontherapy center is under construction.
20. Could there be a combination of nanoparticles (iron or gold) with radiotherapy (protontherapy) in order to increase the efficiency of protontherapy in tumor tissues?
    Research in this direction is in progress to answer that question.
21. Is providing proton and carbon ions using the same system a current trend of equipment manufacturers?
    Yes. Hadron therapy equipment manufacturers are developing systems that enable protontherapy and carbon ion therapy using the same equipment (hybrid systems; e.g., the C400 by IBA).
22. Where can I find data on institutions with hadron therapy equipment?
    The Particle Therapy Co-operative Group (PTCOG) offers highly credible, frequently updated statistics on its website: http://ptcog.ch.
23. What is the penumbra? Can you compare the penumbra for carbon ions and protons?
    The penumbra [80–20 %] is the dose width sideband for the field edge to decrease the dose 80–20 %. [80–20 %] for a collimated beam passively scattered proton varies markedly with depth, being highly dependent on the physical design of the systems that define the beam machine, collimation system, and air space (air gap) between the compensator and the body's surface. Carbon ion therapy has a lower side allowing penumbra a conformational dose laterally and limiting damage to the normal tissue. Protontherapy has a large lateral penumbra, which can cause a higher dose to normal tissue structures in comparison with carbon ion therapy. Therefore, the dose distribution of a single beam appears to be better in carbon ion therapy than in protontherapy.
24. Can the physical and biological differences between proton and carbon ions influence clinical results?
    Carbon ions have greater potential to cause damage to the DNA—in other words, greater isoeffective dose. This, as we know, can be calculated by multiplying the absorbed dose by a weighing factor, which includes radiation quality (LET and RBE). Thus, irreparable damage to the DNA cluster may be greater with carbon ions, as well as apoptosis induction and loss of clonogenicity. Taken together, this can lead to increased cell death.

25. What is the RBE for protons and carbon ions?
Protons have lower RBE, with values in the range of 1.0–1.1. Carbon ions have RBE in the range of 1.5–3.4. The RBE depends on the radiation quality, LET, size fraction, and biological aspects of the target. Uncertainties in RBE can lead to large variations in the current dose released. While a higher RBE is good for tumor control, it is poor for normal tissue toxicity. It has been reported that protontherapy has practical advantages, including lower cost and precise geometric release of beams by use of a gantry. In some centers (HIT, NIRS) of carbon ion therapy, the use of a gantry for operations already exists, so the low-cost advantage only remains for protontherapy.
26. For applications innovative clinics, what is the role of carbon ions versus protons?
The radiobiological properties of carbon ions can improve the local control of tumors and can theoretically replace surgical resection. In addition, carbon ion therapy is indicated for photon-resistant tumors and hard-to-reach places.
27. When deciding to treat with proton or carbon ions, what are the practical considerations?
According to PTCOG data, more than 105,000 patients have been treated with protontherapy; for carbon ion therapy, the number is more than 13,000 patients. The results of the studies established for the comparison of protons and carbon ions are small and difficult to compare. However, the use of carbon ion therapy is recommended when the advantages of using carbon ions outweigh the therapeutic advantages that can already be obtained with fractionated photon RT.
28. What is the role of randomized clinical trials in protontherapy and carbon ions?
Because the RBE of protons is similar to photons, protons can be combined with photons in the course of treatment. Protons are accepted for pediatric treatment; however, for adults there is more controversy due to the high cost of protontherapy. In the literature, one can obtain arguments suggesting that protontherapy and carbon ion therapy cause a reduction in secondary tumors compared to photon therapy. However, it may not be possible to perform a randomized trial to determine this answer, because it would require a large number of patients. At the HIT in Heidelberg, Germany, at least 15 randomized trials are being developed.
29. What is the status of the clinical trials in particle therapy at the University of Heidelberg?
Please see the text, where each test is presented in summary.
30. What is the potential number of patients in the world who would use X-rays, protontherapy, and carbon ion therapy?
Based on studies made in Austria, Germany, France, and Italy in the framework of ENLIGHT (under coordination of Manjit Dosanjh at CERN) and projects in FP7 (ULICE, PARTNER, ENVISION, ENTERVISION), the following estimates can be made (souce: Montarou, G. (2013). Physics and radiobiology in hadrontherapy. Radiobiology in Medicine 17 12 2013):

- X-ray therapy: 20,000 patients/year for every 10 million people
- Protontherapy: 12 % of X-ray patients, or 2,400 patients/year
- Carbon ion therapy for radio-resistant tumors: 3 % of X-ray patients, or 600 patients/year
- *Total:* Approximately 3,000 patients/year for every 10 million people

31. Are there indications of centers of protontherapy closing in the United States? In 2014, it was reported that a protontherapy center was closing in the United States (Indiana University, Protontherapy Center). Probable reasons for this closure include the fact that insurers have started to push back against providing coverage for protontherapy because of a lack of evidence that it provides greater benefits over other treatments, an excess of protontherapy centers in the United States, very old equipment, or administrative problems.
32. What are the current European-fund projects? These projects include the following:

    - ENLIGHT: A wide range of hadron therapy projects (training, research and development, infrastructure) with total funding of ~24 million Euros. All projects are coordinated by CERN, except one coordinated by CNAO).
    - PARTNER: Marie Curie ITN, 12 institutions
    - ENVISION: Research and development on medical imaging for hadron therapy, 16 institutions.
    - ULICE: Infrastructure for hadron therapy, 20 institutions.
    - ENTERVISION: Marie Curie ITN, 12 institutions.

33. What are the 10 most interesting facts about the universe?

    - It was hot when it was young.
    - It will be cold when it grows old.
    - The universe spans a diameter of more than 150 billion light years.
    - It is 13.7 billion years old.
    - Earth is not flat, but the universe is.
    - Large-scale structures of the universe.
    - A huge chunk of it is made up of things we cannot see.
    - There is no such thing as the universe's center.
    - Its members are in a hurry to be as far away from each other as possible.
    - To gain a deeper understanding of it, we need to study structures smaller than the atom.

Index

M. d'Ávila Nunes, *Protontherapy Versus Carbon Ion Therapy*,
Biological and Medical Physics, Biomedical Engineering,
DOI 10.1007/978-3-319-18983-3

MIX
Papier aus verantwortungsvollen Quellen
Paper from responsible sources
FSC® C105338

If you have any concerns about our products,
you can contact us on
ProductSafety@springernature.com

In case Publisher is established outside the EU,
the EU authorized representative is:
Springer Nature Customer Service Center GmbH
Europaplatz 3, 69115 Heidelberg, Germany

Printed by Libri Plureos GmbH
in Hamburg, Germany